# Data and Knowledge in a Changing World


## Scientific Editorial Board

E. Fluck (Chairman) (*Germany*)
A. Kolaskar (*India*)
K.W. Reynard (*U.K.*)
C.-H. Sun (*China*)
M.A. Chinnery (*USA*)
J.-E. Dubois (*France*)
P. Glaeser (*France*)
F.A. Kuznetsov (*Russian Federation*)
G.H. Wood (*USA*)

## Corresponding Members

D. Abir (*Israel*)
M. Antoninetti (*Italy*)
M. Attimonelli (*Italy*)
C. Bardinet (*France*)
C. Bernard (*France*)
H. Bestougeff (*France*)
Y. Bienvenu (*France*)
F.A. Bisby (*U.K.*)
L. Blaine (*USA*)
G. Blake (*France*)
J.P. Caliste (*France*)
M. Chastrette (*France*)
E.R. Cohen (*USA*)
M.-T. Cohen-Adad (*France*)
A.-F. Cutting-Decelle (*France*)
M. Delamar (*France*)
J.-L. Delcroix (*France*)
R. Eckermann (*Germany*)
N. Gershon (*USA*)
E.R. Hilf (*Germany*)
S. Iwata (*Japan*)
J.G. Kaufman (*USA*)
H. Kehiaian (*France*)
A.D. Kozlov (*Russia*)
H. Lalieu (*Netherlands*)
D. Lide (*USA*)
P. Masclet (*France*)
J.C. Menaut (*France*)
B. Messabih (*Algeria*)
P. Mezey (*Canada*)
F. Murtagh (*Germany*)
G. Ostberg (*Sweden*)
W. Pillman (*Austria*)
J.R. Rodgers (*Canada*)
S.F. Rossouw (*South Africa*)
J.J. Royer (*France*)
J. Rumble (*USA*)
S.K. Saxema (*Sweden*)
R. Sinding-Larsen (*Norway*)
R.J. Simpson (*Australia*)
N. Swindells (*U.K.*)
M. Tasumi (*Japan*)
B.N. Taylor (*USA*)
J. Thiemann (*USA*)
A. Truyol (*France*)
A. Tsugita (*Japan*)
P. Uhlir (*USA*)
F. Webster (*USA*)
E.F. Westrum (*USA*)
G. Wood (*Canada*)
Z. Xu (*China*)

Springer

*Berlin*
*Heidelberg*
*New York*
*Barcelona*
*Budapest*
*Hong Kong*
*London*
*Milan*
*Paris*
*Santa Clara*
*Singapore*
*Tokyo*

J. O. Dubois · A. Gvishiani

# Dynamic Systems and Dynamic Classification Problems in Geophysical Applications

With 76 Figures and 4 Colour Plates

Springer

CODATA

CODATA Secretariat
Phyllis Glaeser, Exec. Director
51, Boulevard de Montmorency
F - 75016 Paris

Authors:

Prof. Dr. Jacques Octave Dubois
Institut de Physique
du Globe de Paris
B89, 4 place Jussieu
75252 Paris Cedex 05
France
*e-mail: dubois@ipgp.jussieu.fr*

Prof. Dr. Alexei Gvishiani
Centre of Geophysical Data Studies
and Telematics Applications
Molodezhnaya Str., 3
117 296 Moscow
Russian Federation
*e-mail: gvi@wdcb.rssi.ru*

*The image on the front cover comes from an animation which shows world-wide Internet traffic. The color and height of the arcs between the countries encode the data-packet counts and destinations, while the "skyscraper" glyphs (or icons) encode total traffic volume at any site. This image was generated by Stephen G. Eick at the AT&T Bell Laboratories.*

ISBN 978-3-642-49953-1 ISBN 978-3-642-49951-7 (eBook)
DOI 10.1007/978-3-642-49951-7

Cataloging-in-publication Data applied for

Die Deutsche Bibliothek - Cip-Einheitsaufnahme
**Dubois, Jacques O.:**
Dynamic systems and dynamic classification problems in geophysical applications /
J. O. Dubois ; A. Gvishiani. ICSU CODATA. - Berlin ; Heidelberg ; New York ;
Barcelona ; Budapest ; Hong Kong ; London ; Milan ; Paris ; Santa Clara ; Singapore ;
Tokyo : Springer, 1998
(Data and knowledge in a changing world)



Softcover reprint of the hardcover 1st edition 1998



Typesetting: Camera-ready by editors
SPIN:10555332 51/3020-5 4 3 2 1 0 - Printed on acid-free paper

# Foreword

This book is the latest volume in the series entitled " *Data and Knowledge in a Changing World* ", published by the Committee on Data for Science and Technology (CODATA) of the International Council of Scientific Unions (ICSU). This series was established to collect together, from many diverse fields, the wealth of information pertaining to the intelligent exploitation of data in the conduct of science and technology.

This volume is the first in a two-volume series that will discuss techniques for the analysis of natural dynamic systems, and their applications to a variety of geophysical problems. The present volume lays out the theoretical foundations for these techniques. The second volume will use these techniques in applications to fields such as seismology, geodynamics, geoelectricity, geomagnetism, aeromagnetics, topography and bathymetry.

The book consists of two parts, which describe two complementary approaches to the analysis of natural systems. The first, written by A. Gvishiani, deals with dynamic pattern recognition. It lays out the mathematical

theory and the formalized algorithms that forms the basis for the classification of vector objects and the use of this classification in the study of dynamical systems, with particular emphasis on the prediction of system behavior in space and time. It discusses the construction of classification schemes, and the evaluation of their stability and reliability.
The second, written by J. Dubois, is concerned with various theoretical tools that may be applied to the modeling of natural systems using large sets of geophysical data. Dynamic systems are characterized in terms of attractors and repellers, and methods for geometrical classification are described. These are integrated together with elements of chaos theory into the concept of self-organized criticality, and this used to describe the evolution of dynamic systems.
While most applications are to be discussed in the second volume, a few illustrations are included in the present book, to demonstrate the need for the analytical tools. These are drawn from the field of seismic hazard assessment, and the use of classification schemes in the determination of earthquake-prone areas. These methods have been applied in a number of geographical areas with considerable success.

The book is expected to be of interest to scientists, engineers, applied mathematicians and computer programmers who use artificial intelligence methods in the analysis of large data sets. The authors are particularly concerned with geophysical data sets, but the techniques described will have applications in many other fields. It will be of particular interest to those concerned with the prediction of different kinds of natural processes that develop in space and time, and with the monitoring and testing of various prediction algorithms.

Dr. Michael A. Chinnery
CODATA
Boulder, Colorado 80303-3526, U.S.A.

# Table of Contents

# Part III
# Dynamic Systems

# Acknowledgement

This book presents the results of more than 15 years of intensive cooperation between French and Russian (Soviet) geophysicists, mathematicians and informaticians in the fields of non-linear mathematics, artificial intelligence and geoinformatics applied to geophysical and natural hazard data analysis. Working on the book, the authors maintained permanent collaboration with major international organizations dealing with global data managment matters such as CODATA, ICSU Panel on World Data Centers, European-Mediterranean Seismological Centre and Coordinating Committee on Data Exchange and Centres of the International Lithosphers Program.

Professor Jacques Émile Dubois, the President of CODATA encouradged the authors to orient this study towards CODATA goals and objectives what gave a real opportunity to present this book in the series "DATA AND KNOWLEDGE IN A CHANGING WORLD".

On different stages of the project, joint French and Russian working group of mathematicians, informaticians, geophysicists and geologists contributed to development and evaluation of presenteded theories and algorithms, as well as to their applications to geophysical data analysis.
The French contributors were: Georges Jobert, Claude Jean Allègre, Jean-Louis Le Mouël, Vincent Courtillot, Raul Madariaga, Jean-Paul Montagner, Michel Diament, Pascal Bernard (Institut de Physique du Globe de Paris), Jean Bonnin, Armando Cisternas, Michel Cara, Daniel Rouland (Institut de Physique du Globe de Strasbourg), Jean Sallantin, Hervé Philip (Université de Montpellier), Christian Weber (BRGM), Bagher Mohammadioun (CEA).
The main Russian (Soviet) contributors were: Alexander Ianshin (Institute of Lithosphere, Moscow), Dimitri Rundkuist (Institute of Geology and Geochronology of St Petersburg), Michael Sadovsky, Vladimir Strakhov,

Vladimir Keilis-Borok, Alexander Soloviev, Vladimir Gurvitch, Vladimir Kossobokov, Anatoli Levshin, Michael Zhizhin, Alexander Troussov (Institute of Physics of the Earth, Moscow).
All of them are here greatly acknowledged.

We acknowledge the really important contribution of French CNRS and INSU (respectively Centre National de la Recherche Scientifique et Institut National des Sciences de l'Univers) and the Institut de Physique du Globe de Paris that provided continuous support to this study by organizing and funding visits of Russian (Soviet) scientists in France. We are also grateful to directors of IPGP, Jean-Louis Le Mouël and Vincent Courtillot, to Michel Diament director of the Laboratoire de Gravimétrie et de Géodynamique, to Guy Aubert director of foreign affairs of IPGP, as well as to officers of these institutions who provided excellent administrative and coordinating job: Jeanine Mivielle (IPGP head of the personal)and Liliane Flabbée secretary general of IPGP. Antoine Sempere, head of CNRS office in Moscow is warmly acknowledged for his permanent efficient help of the project development. We acknowledge as well the important contribution of Joint Institute of Physics of the Earth in Moscow, headed by Vladimir Strakhov, National Geophysical Committee of Russian Federation headed by Gennadi Sobolev and administrative officers of Russian Academy of Sciences.

We are also grateful to Mrs Bernice Dubois who corrected English language of the manuscript, Dr. Michael A. Chinnery who accepted to write the foreword, Steve Pride who reviewed the manuscript in its totality, Hélène Robic who solved the main problems in presentation of figures and of the final form of the manuscript, Jean Jacques Royer for the cover drawing, Alexander Beriozko and David Aubert for preparation of computerized versions of some figures.

# Introduction

This book is devoted to nonlinear dynamics and dynamic pattern recognition applied to Geophysical Data Systems in seismology, geodynamics, geomorphology, geomagnetism and natural hazard assessment. It describes the theoretical and algorithmic results of more than ten years of studies in dynamic systems and artificial intelligence applied to geophysical problems. Much in this book is a result of Franco-Soviet (1980-1991) and then Franco-Russian (1991-1996) collaboration in mathematical geophysics. Both authors were involved in this collaboration from the beginning. The results have been published in many international, French and Russian scientific journals. Two books have been published on the topic in Russian [81] and one in French [44].

There are two general approaches to investigate any applied (in particular, geophysical) problem. The first is to construct many detailed physical and mathematical models of the process. This leads to non-linearity pitfalls in the process under consideration which are only seldom adequately reflected in the mathematical equations. The other approach is to model underlying features process and to try to recognize its main features in the phase space of the available data sets. This book deals with the second approach, developing it on the basis of dynamic systems and classification methods and algorithms of pattern recognition with learning.

The book consists of two main parts. They are both written in rather mathematical language, and present theoretical aspects of the dynamic methods introduced. Geophysical applications of mathematical and informatics techniques described in the book will be the subject of the next book which the authors also plan to write for publication in the CODATA series *"Data and Knowledge in a Changing World"*.

At the moment the number of such studies in the geosciences is increasing rapidly. The common feature of the applications is to long term data series. Discrete dynamic systems and models of dynamic pattern recognition problems are well adapted to represent these data and their properties, thereby providing a new approach to understand their behaviour.

In seismology and seismic-hazard assessment, tectonically and seismically active areas are presented as dynamic systems. Thus, the interpretation of the seismic phase space flow is generally possible because of the extensive earthquake catalogues. Scaling laws applied in space, time and energy clearly imply Self Organized Criticality (S.O.C.) processes.

After recalling some basic mathematics, the first part of the book is devoted to the theory and algorithms of dynamic pattern recognition problems and their use in controlling the reliability of classifications in discrete metric spaces. The importance of relevant applications (such as natural hazard assessment problems) requires special conditions satified by the mathematical classification tools used. Furthermore, if an applied problem under consideration requires a very precise solution, then the question of classification quality control becomes prominent. An example of such a problem is the recognition of strong earthquake-prone areas in regions of high and moderate seismicity. This problem in applications is considered shortly in the Chapter 1 of the book.

The price of an error in this kind of problem can be very high. Indeed, the planning of cities, towns, nuclear power plants, chemical factories, transportation systems, life lines, etc., depend on correct identification of strong earthquake-prone areas in the region.

Following this line, new types (classes) of dynamic pattern recognition problems are introduced. To solve them, a new family of algorithms with learning " Voting by a set of features" (VSF) is developped. Dynamic pattern recognition problems include classical (static) classification problems as a particular case.

For dynamic pattern recognition problems, we introduce the condition of stability, considered necessary for accepting a classification. Appropriate mathematical theorems are tested for verification of this condition.

The second part of the book describes general theoretical tools addressing the study of natural dynamic systems. The tools are based on the notions of Hausdorff measure and dimension. The continuous and discrete forms fot characterizing dynamic systems are then described. Discrete dynamic systems are defined by use of the Iterate Function Scheme analysis. This permits an easy introduction of the foundations of dynamic systems studies – attractors and repellers, bifurcations in mapping processes due to variations of the control parameters, Lyapunov exponents, etc. The dynamic systems theory is presented in a form which will be easily readable by a wide range of geophysicists.

The study focuses on the geometry of attractors using a correlation function test, first return mapping, Lyapunov exponent computation, etc. After such geometrical classification we return to bifurcations and harmonic cascades to introduce the concepts of chaos, the road to chaos and the control of chaos.

Finally, we define Self Organized Criticality (S.O.C.) which synthesizes the above constructions. Actually, power laws of different kinds enter simultaneously into the evolution of the dynamic systems as into natural ones, *e.g.* earthquake swarms, volcanic eruptions, or the geomagnetic dynamo.

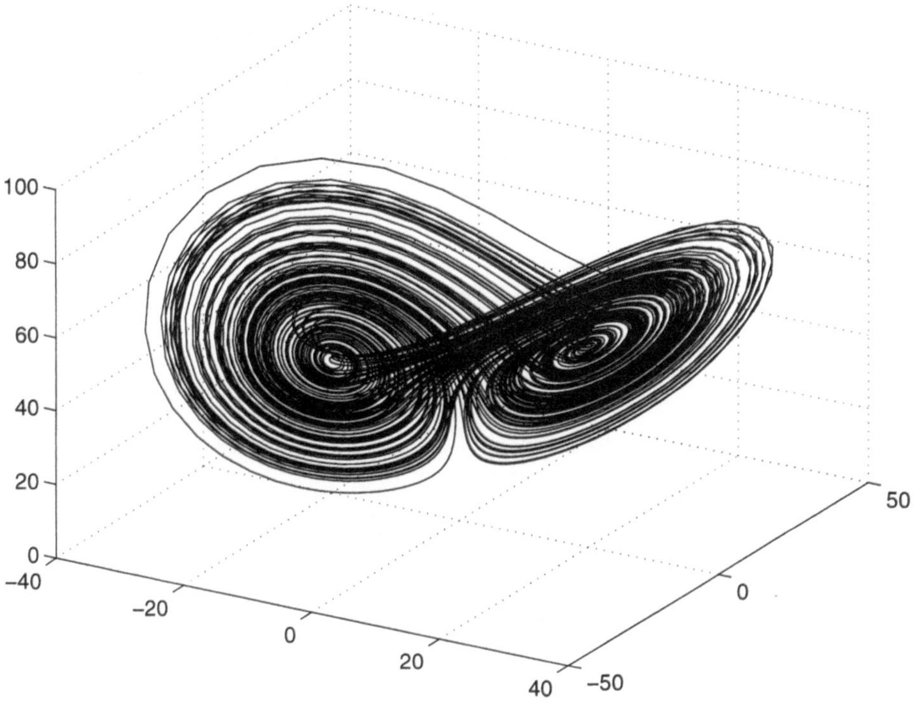


The Lorenz Attractor – The parameters (see the text) are: $Pr = 10$ ; $b = 8/3$ ; $r = 45.92$ The perspective representation in the phase space shows 10,000 points discrete phase trajectories (after David AUBERT, 1997).

# Chapter 1

# Why the Techniques Developed in this Book are Important? (A Few Examples of Applications)

The techniques described in this book look rather complicated at first glance and may seem too long to realize for those who deal with classification studies of real geophysical data (or any other kind of data to which the techniques are applicable). Indeed, this is the price (and a rather reasonable one) for the fact that by use of the above techniques we pursue the ultimate goal of a classification study – to classify the data following the rules given by learning material and to choose the most reliable classification among those which may be accepted according to different necessary conditions.
To show the reader the attractiveness of the techniques, we give herein just a few vivid examples of results obtained by the methods presented in this book.
Prediction of strong earthquake-prone areas is an interesting field of application of dynamic pattern recognition techniques (see 6). Furthermore, any kind of natural disaster spatial prediction for a process which develops in time must be tackled by dynamic classification algorithms and a dynamic systems approach.

The results of concrete geophysical and natural hazard assessment applications of the techniques developed in this book will be studied in another book comprising the second part of this study. Important applications of dynamic pattern recognition and dynamic systems techniques as well as dynamic programming of seismological and geodynamic data are on their way, in particular in the framework of a joint research programme between the Institute of Physics of the Earth in Paris (Institut de Physique du Globe de Paris, IPGP) and the Center of Geophysical Data Studies and Telematics Applications of the Schmidt Joint Institute of Physics of the Earth in Moscow. Therefore the authors consider that this second book should appear in 1-2 years from now.

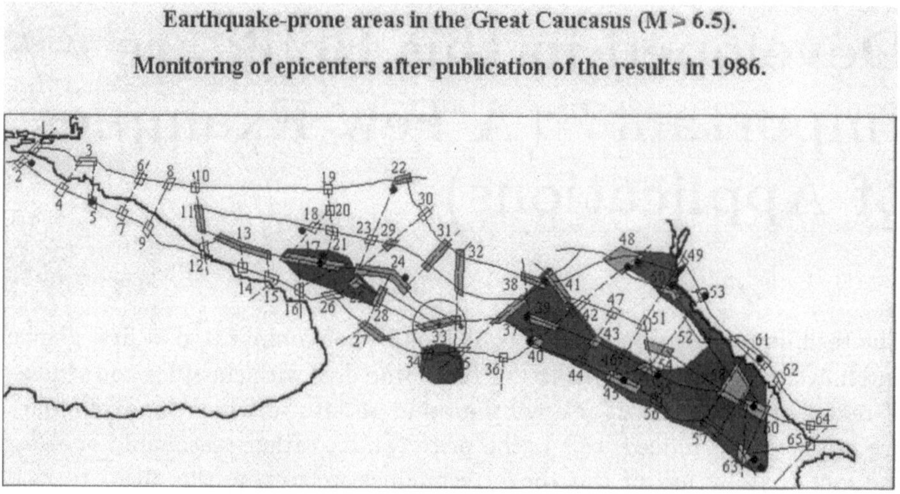


Figure 1.1: *Successful real-time prediction of the place of the Racha-Java earthquake 04 29 1991 in Georgia, the Great Caucasus.* 1. The grey rectangles are the dangerous knots for the magnitude $M \geq 6.5$ recognized by a dynamic pattern recognition algorithm, and published in 1988. 2. Knot 33 surrounded by a circle is the location of the disastrous Racha-Java earthquake 04 29 1991, M = 6,9 in Georgia. See the text for more information.

*(see also colour plate on page 254)*

In figures 1.1, 1.2 and 1.3 we see the results obtained by dynamic pattern recognition techniques in identifying earthquake-prone areas in three regions of moderate seismicity: the Great Caucasus (fig. 1.1), the Western Alps (fig. 1.2) and the Pyrénées (fig. 1.3). More than 15 years have passed since

publication of the results and we were able to compare these with actual seismic events throughout this period of time.
The results of classifying disjunctive knots in the Caucasus, sections of tectonic faults in the Alps and intersections of morphostructural lineaments in the Pyrénées are shown in figures 1.1, 1.2 and 1.3.

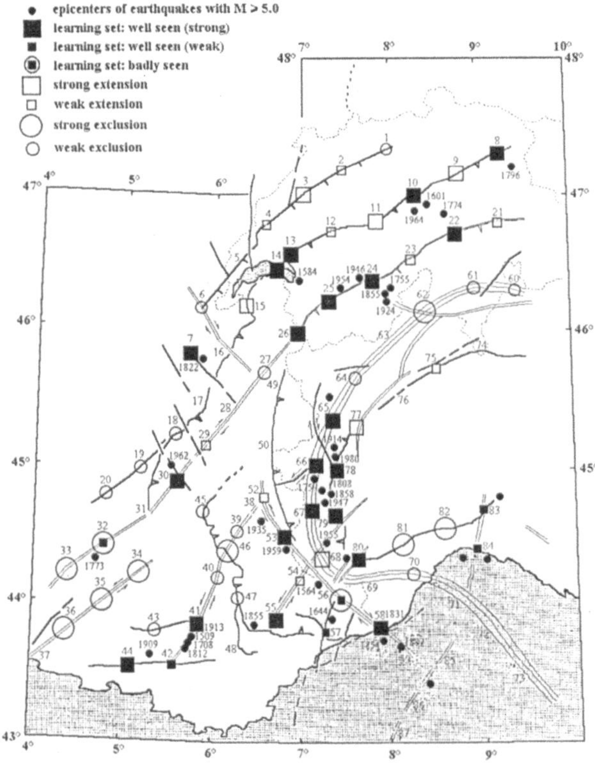


Figure 1.2: *Identification of strong earthquake-prone areas in the Western Alps by dynamic pattern recognition techniques.* 1. Large black and white squares – the most seismically dangerous ($M \geq 5.0$) segments of tectonic faults in the Western Alps identified by the Expert Communication Algorithm. 2. Large and small white circles are the least dangerous segments of the faults on the ? scheme of the Western Alps. 3. Small black circles are the epicenters of known earthquakes with $M \geq 5.0$ in the Western Alps.

We observe the most important fact in the Great Caucasus. The site of Racha-Java where a disastrous earthquake occurred on April 29th, 1991 in Georgia, was recognized three years in advance (see figure 1.1). The corresponding result was published [81].
The classification of these three regions as the areas of moderate seismic-

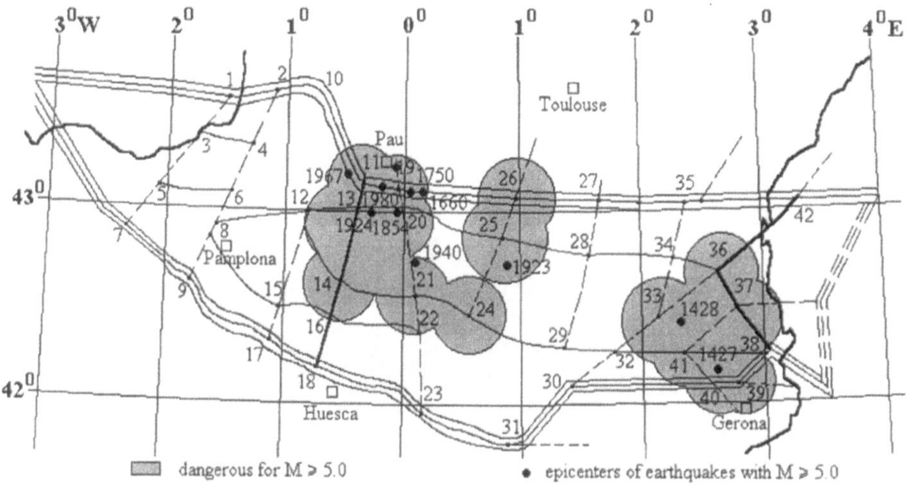


Figure 1.3: *Strong Earthquake-Prone Areas in the Pyrénées ($M \geq 5.0$) recognized by dynamic pattern recognition techniques.* 1. Gray zones are identified as earthquake prone with $M \geq 5.0$ by dynamic pattern recognition techniques. They extend two known zones of seismic activity in the Pyrénées. 2. Small black circles are the epicenters of known earthquakes with $M \geq 5.0$ in the Pyrénées. 3. The lines are the morphostructural lineaments.

ity is not quite so strict. In all three mountain countries historical earthquakes with epicenter intensity I = X are known. At the same time the level of recent seismicity is significantly lower than for [66, 67, 68, 69, 73], where pattern recognition CORA-algorithm was used for determination of earthquake-prone areas for the large magnitude earthquakes. While studying the earthquake-prone areas in the Great Caucasus, the Alps and the Pyrénées, we realized that more sophisticated techniques based on dynamic pattern recognition algorithms were needed. Such techniques were developed and applied to studies of earthquake-prone areas in regions of moderate seismicity.

The studies of potentially seismic zones by pattern recognition techniques were started in the early 70s. In the first half of the 80s attention in this field was concentrated on controlling reliability of the results. Since input information was still rather limited, this necessarily led to the application of mathematical methods such as geometry in discrete metric spaces, combinatorics, metric space theory, game theory, mathematical statistics and elements of functional analysis.

Using these methods, A. Gvishiani and V. Gurvitch showed in [83] that the recognition of earthquake-prone areas is a special case of a pattern recognition problem called by the authors a dynamic classification problem. The formal mathematical study of this construction, given in this book, leads to the formulation of the stability condition that allows one to distinguish a class of the most reliable predictions. This is the main goal of the theory presented in this book.

We give examples of such stable predictions of strong earthquake-prone areas in the great Caucasus (fig. 1.1), the Western Alps (fig. 1.2) and the Pyrénées (fig. 1.3).

We include these examples in a theoretical book to provide the reader with "samples" of the final products that can be obtained by the dynamic pattern recognition and dynamic systems techniques.

We comment briefly on the results presented in figures 1.1-1.3. The earthquakes whose sites we want to predict have thresholds of magnitude $M \geq 6.5$ in the Great Caucasus (fig. 1.1) and $M \geq 5.0$ in the Western Alps and the Pyrénées.

In the case of the Great Caucasus the objects of recognition are disjunctive (morphostructural) knots (see [81] that surround intersections of morphostructural lineaments presented in figure 1.1. The epicenters of known strong earthquakes ($M \geq 5.5$) are plotted on the figure as black dots. The knot N33 located in the central-southern part of the figure, (where strong earthquakes were not known before) is recognized as dangerous for potential occurrence of earthquakes with magnitude $M \geq 6.5$ in [81] published in 1987. The Racha-Java earthquake occurred on April 29th 1991. It was the strongest recent earthquake on Georgian territory. It was a piece of luck that the epicenter of this earthquake located in the predicted seismically dangerous knot N33 occurred sufficiently far away from big cities and important economic sites. Otherwise the damage produced might have been comparable with the disastrous Spitak earthquake in Armenia [32], where more than 24 000 people were killed. As the reader can see in figure 1.1, both the main shock ($M = 6.9$) and the strongest aftershock ($M = 6.3$) are located inside the knot N33 that was recognized as dangerous for earthquakes $M \geq 6.5$ [81] as well as for earthquakes with $M \geq 5.5$. Since strong earthquakes were not previously known in this spot, this provides strong evidence in favor of the result. Furthermore, in the whole system of regions [66, 70, 73, 79, 80, 81, 82, 87, 89, 90, 92] where pattern recognition of strong earthquake-prone areas was executed, more than 25 cases are registered where epicenters of the strong earthquakes (which occurred after publication of the corresponding papers) are located within the sites recognized as dangerous. This is obviously important evidence in favor of applying the technique developed in this book to natural hazard assessment. Among these earthquakes is the disastrous Spitak quake referred to above that occurred on December 7 1988 in Armenia ( $M = 7.0$ epicentral intensity $I = 9 - 10$) [32] which produced major damage to important economic sites and killed more than 24 000 people. The corresponding knot was classified for $M \geq 6.5$ as dangerous in [90].

In figure 1.2 the objects of recognition are the fault segments of the neotectonic scheme of the Western Alps constructed by C. Weber and his group [31]. The classification was obtained by the Expert Communication Algorithm introduced by M. Sallantin. The large black squares (learning set: very visible) and wide transparent squares (strong extension) are those parts of the faults in whose vicinity earthquakes with $M \geq 5.0$ may occur.

In the case of the Pyrénées, we have another type of recognition objects. There are the intersections of morphostructural lineaments (fig. 1.3). Again, black dots are the epicenters of earthquakes with $M \geq 5.0$. The dark gray zones are those recognized as seismically dangerous for earthquakes with $M \geq 5.0$. As we see in the figure the two recognized gray zones extend two known zones of strong historical and instrumental seismicity in the Northern-Central and the South-Eastern parts of the Pyrénées. This extension provides an important added value to the official maps of seismic zoning of France and Spain[1].

The next important impact in developing dynamic pattern recognition techniques and dynamic programming applications to geophysical data analysis came in 1990-1995. At that time the largely new idea of syntactic representation of recognition objects was introduced by J. Bonnin, A. Gvishiani and M. Zhizhin [93]. Incorporating the ideas described in this book and a syntactic pattern recognition approach (derived from speech recognition applications), they designed the new classification algorithm SPARS (Syntactic Pattern Recognition Scheme). This algorithm was successfuly applied by A. Gvishiani, M. Zhizhin, J. Bonnin, B. Mohammadioun, R. Madariaga and D. Rouland to the analysis and classification of waveforms [93] and by M. Zhizhin, J. Battaglia, J. Dubois and A. Gvishiani to the analysis of geodynamic modeling of the ocean crust spreading [186].
The main ideas of this algorithm are illustrated in figure 1.3. Where we need to compare scalar or vector waveforms of time signals, SPARS serves as a tool which gives more efficient results than are obtained with the correlation generally used for such a comparison. SPARS techniques have been successfully applied to classification of earthquake accelerograms according to tectonic regions and to recognition of the earthquake epicenter clustering in the Vanuatu region on the basis of seismogram analysis.
Concerning SPARS applications, a reconstruction of the dynamics of the lithospheric sea-floor spreading in the Mid Atlantic Ridge area can be described on the basis of magnetic anomaly profiling [186]. The algorithm's input was a set of magnetic anomaly profiles orthogonal to the ridge axis, measured with a fixed step, compared to a theoretical synthetic profile com-

---

[1]The pattern recognition study of the Great Caucasus, the Western Alps and the Pyrénées was done in the framework of a collaboration project between the Moscow Institute of Physics of the Earth and the Moscow Institute of Geography on the Russian (Soviet at that time) side and les Instituts de Physique du Globe de Paris et Strasbourg and l'Université de Montpellier on the French side. Many well known French and Russian scientists were involved in the project. Among them are: A. Cisternas, H. Philip, J. Bonnin, A. Soldano, J. Sallantin, P. Godefroy, C. Weber, J. Lambert, M. Diament, V. Keilis-Borok, V. Kossobokov, A. Soloviev, A. Gorshkov, M. Zhidkov, E. Rantzman, A. Troussov, V. Gourvitch as well as the authors of this book. The project gained important support from the French Centre National de la Recherche Scientifique (CNRS) and from the Soviet Academy of Sciences, which organized in 1980-1986 the joint French-Soviet working group on the project. The group was co-chaired by G. Jobert and then by J. Dubois on the French side and by M. Sadovsky (vigourous head of Russian Geophysics during more than 20 years) on the Soviet side. A. Gvishiani was vice-chairman of the working group from the Russian side. C.J. Allègre played an active role in the discussions of this group.

puted on a simplified model of Earth crust accretion with a constant spreading rate. Figure 1.4 clearly shows the differences between the SPARS results and that obtained by the cross-correlation method. SPARS allows us to identify many more details and in particular to date the crust in each point of observation and to study time-space variations of the velocity field of the ocean floor spreading.

Once again, in this chapter, we first mention very briefly the applications obtained by the techniques described below.

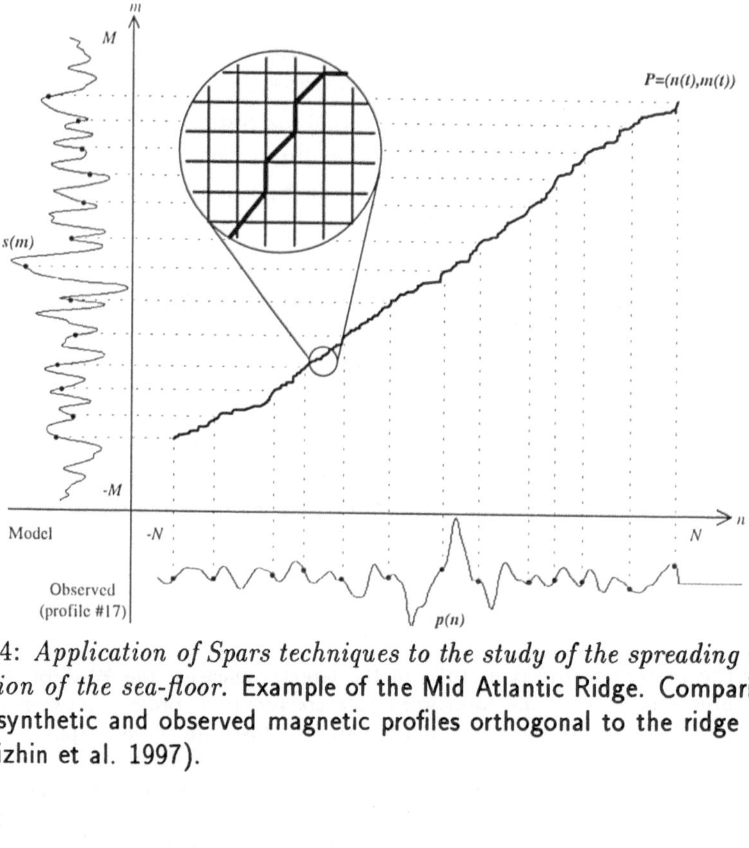


Figure 1.4: *Application of Spars techniques to the study of the spreading rate of accretion of the sea-floor.* Example of the Mid Atlantic Ridge. Comparison between synthetic and observed magnetic profiles orthogonal to the ridge axis (after Zhizhin et al. 1997).

With regards to dynamic systems approach, we shall limit ourselves to two domains: fracturation and seismicity for the first and geomagnetism and geodynamo for the second.

Applying the Cantor dust method to seismic time series and to volcanic eruption series it has been shown [165, 47, 45] that seismic and eruption sequences are not distributed randomly but as power law. The same type of distribution was observed long ago by Gutenberg and Richter for seismic energy dissipation [78].

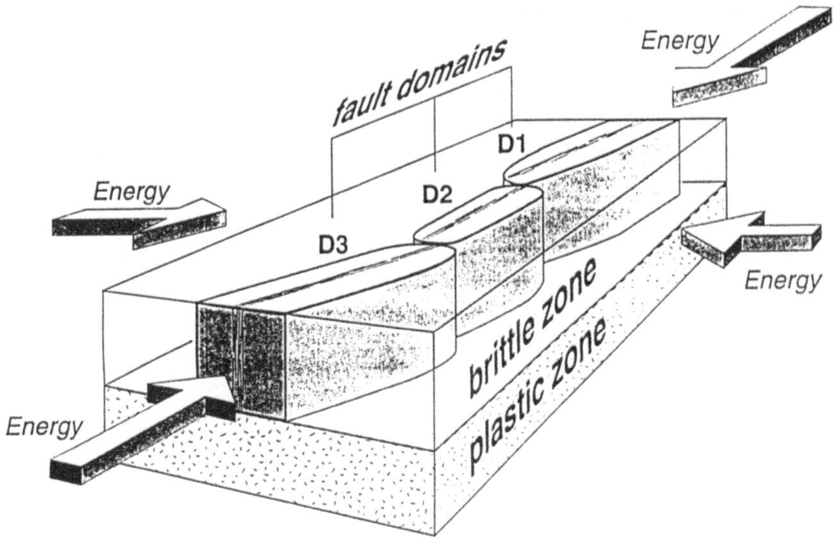


Figure 1.5: *SOFT model, the geometry of a certain plate tectonics boundary zone.* The fault zone is made of different segments with different geometries. The fault zone is supposed to be composed of a brittle layer above a plastic one. The areas receive energy continuously from plate tectonic movement (see Allègre *et al.* 1995).

In their theoretical approaches Allègre *et al.* and Blanter *et al.* [3, 4, 5, 21], starting from models of fracturation based on the renormalized group method, were able to reproduce the time sequences of seismic crises and of their energy dissipation. The methods presented in SOFT models are links between physical methods, e.g. multiblock models (see the Burridge and Knopoff model of theoretical seismicity [30]) and scaling law distribution of earthquake occurrence time series.
These methods will be described in detail in the second book devoted to applications. Here we only give an illustration of this approach. In one of their models Allègre *et al.* (1995) [5] consider a fault zone receiving a certain elastic energy resulting from the stress applied to its boundary by plate tectonics as shown in figure 1.5
In figure 1.5 three arbitrary domains were chosen. The fault zone is supposed to be composed of a brittle layer above a plastic one. Around each segment a three-dimensional domain is defined with specific geometries extending to a prescribed depth. The three-dimensional domains are those in which the fractures will occur. The quantum of energy from plate tectonic origin

is assumed to be constant in a unit of time. When the energy per unit of surface is larger than a certain quantity, the excess energy can be used to generate new cracks or develop ancient cracks. The authors use scaling techniques as did Allègre *et al.* (1982) [3] and Allègre and Le Mouël (1994) [4] to compute the probability of cracks at scale 2 when the probability of cracks is known at the elementary scale 1, according to the renormalized group method.

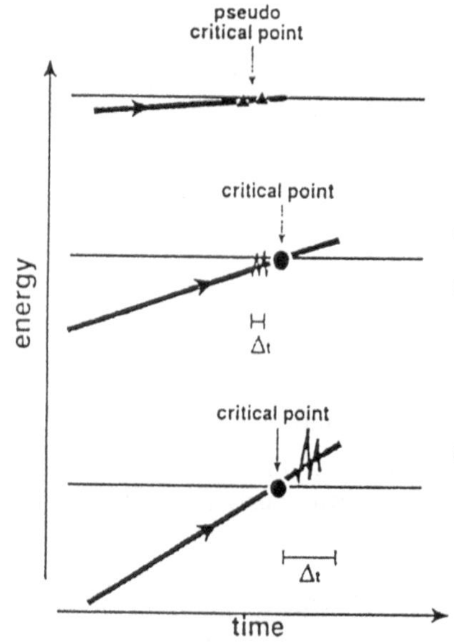


Figure 1.6: *SOFT model, the behavior of fault domains according to the energy rate.* The three different responses according to energy input are represented. The horizontal black line represents the critical probability (or critical energy per surface unit). When critical energy is approached slowly the response is a continuous loss of energy without any real critical phenomena. When the energy increases more quickly, foreshocks precede the main shock followed by aftershocks. When energy increase is steeper a big earthquake occurs without precursors (Allègre *et al.* 1995).

The numerical experiment gives interesting results which are detailed in figure 1.6. The horizontal black line is the critical probability or critical energy per surface unit. If the energy is approached slowly by the rise of elastic energy, the response is a continuous loss of energy without any real critical phenomena. If the energy increases more quickly, foreshocks (pseudocritical phenomena) lead to a true critical point, a strong earthquake.

For a steeper increase of energy, the system passes the critical point and a strong earthquake occurs directly, without precursors

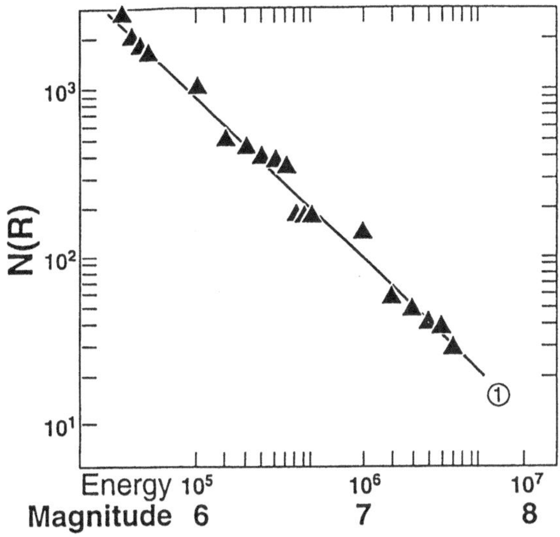


Figure 1.7: *Theoretical Gutenberg-Richter law in the SOFT model results.* This diagram was obtained from sommation of different experiments on the SOFT model. (Allègre *et al.* 1995).

The last point, which shows the very coherent results of such a model, deals with the Gutenberg-Richter statistics. Let us recall that the Gutenberg-Richter law on statistics of the magnitudes of earthquakes appears to be a universal scaling law in seismology [78]. In order to check whether the model satisfies it, the authors built a number of series earthquakes. They computed the cumulated histogram numbers of shocks versus their magnitude and drew the corresponding log-log diagram. In figure 1.7 we observe the good fitting of the power law.

Let us now consider examples of theoretical models of earth geomagnetic field instabilities. It is well known from paleomagnetic studies that the dipole component of the Earth's magnetic field reverses irregularly. Cox (1968) [35] proposed that the interval between two successive reversals is a random quantity governed by a Poisson distribution with a mean of about one million years. As mentioned by Ershov *et al.* (1989) [53], it is, however, unrealistic to introduce randomness into the large-scale dynamo of the Earth. According to modern ideas, chaos in dynamics can arise naturally from the nonlinearity of a deterministic system (see *e.g.* Guckenheimer and Holmes, 1983, [77]).

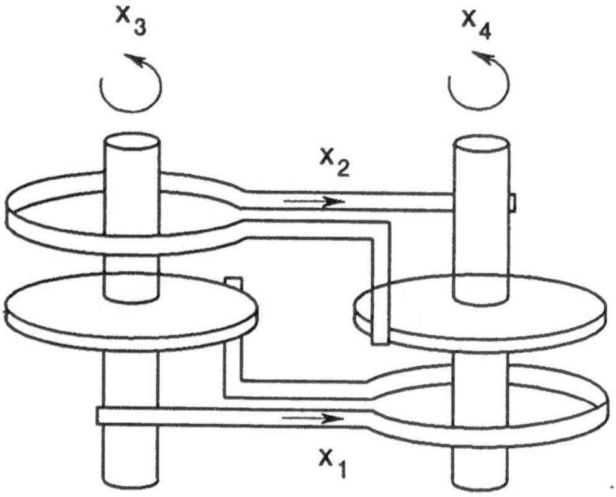


Figure 1.8: *The two-disk dynamo.* The model is given by Rikitake (1958), Ito (1980) Ershov *et al.* (1989). It indicates that $x_1$ et $x_2$ are dimensionless measures of the currents in the loops, while $x_3$ and $x_4$ are proportional to the angular velocities of the disks.

The Rikitake system (Rikitake, 1958, [150]) is a model to explain the reversal of the Earth's magnetic field. It consists of coupled, self-excited disc dynamos (figure 1.8. Equal constant torques act on their shafts. Adding viscous friction that acts to reduce the angular momentum of the disks, we obtain the system of equations

$$\begin{cases} \dot{x_1} &= -\mu x_1 + x_2 x_3, \\ \dot{x_2} &= -\mu x_2 + x_1 x_4, \\ \dot{x_3} &= 1 - x_1 x_2 - \nu_1 x_3, \\ \dot{x_4} &= 1 - x_1 x_2 - \nu_2 x_4 \end{cases} \tag{1.1}$$

where variables overlined by a dot indicate their time derivative. These variables, in non-dimensional form, $x_1$ and $x_2$ are the electric currents in the disks, $x_3$ and $x_4$ are their angular velocities, $\mu$ is the ohmic dissilation coefficient, the same for both circuits, and $\nu_1$ et $\nu_2$ are the coefficients of viscous friction which may be different for each disk.
In both studies on the Rikitake model by Ito (1980) and by Ershov *et al.*

(1989) numerical integration of the differential equation system and representation in the phase space gives attractors very similar to the Lorenz type, characterized by irregular travelling of an orbit between two unstable fixed points. Travelling corresponds to polarity reversal. The frequency of polarity reversals depends strongly on the parameter $\mu$ the ohmic dissipation coefficient in the core.

In the more recent study by Hide (1995) [102] and Hide *et al.* (1996) [103], two novel self-exciting single-disk homopolar dynamo models were proposed. Similar deterministic chaotic behaviour was observed by Dubois and Pambrun (1990) [50] in their study on distribution of reversals of Earth's magnetic field from 165 Ma to the present time. They considered the time distribution of the 296 reversals of the earth's magnetic field in Cox's scale. They observed that an exponential distribution describes the set of reversals that occur at intervals shorter than 0.5 Ma. For longer ones the distribution is found to fit a power law. They examined the sequence in terms of deterministic behaviour and singled out an Attractor in a phase space, where the variable is the timelength of direct or inverse polarity. Using a sliding window technique the attractor was found to be close to 2, implying 2 degrees of freedom for the process which generates the reversals. This process appeared to be polluted by "external" noise during the intervals 165 to 140 Ma and 23 Ma to the present time (see figure 1.9).

A controversial debate on this problem arose when Marzocchi (1997), Marzocchi and Mulargia (1992) and Marzocchi *et al.* (1995) [131, 132, 133] emphasized, in studying the periodicity of geomagnetic reversals, that the results indicate that geomagnetic reversals since 85 Ma BP occur according to a generalized Poisson process with a mean which increases exponentially, and which may have a superimposed periodic modulation [132]. They concluded that statistical comparison of synthetic series of geomagnetic reversals with the real series of reversal indicates that the model is inadequate to represent reality [133].

In another analysis on a more complete reversal time series Cortini and Barton (1994) [33] concluded that, because of the small size of the geomagnetic reversal data set (only 282), and the poor definition of the scaling region, the correlation dimension for the magnetic period sequence is quantitatively not meaningful. The clear difference in correlation integrals between measured and randomized reversal sequences suggests that the geomagnetic reversal dynamics is not random and that low-dimensional chaos, if not detected, can be suspected. We shall stop at this stage, but a more detailed analysis on this very important point will be developed in our second book devoted to applications of our nonlinear analysis tools.

The authors hope that in the coming years great progress will be made in the field of artificial intelligence methods applied to geoscience data analysis. In this approach, dynamic pattern recognition, dynamic systems approach and dynamic programming will play an important role.

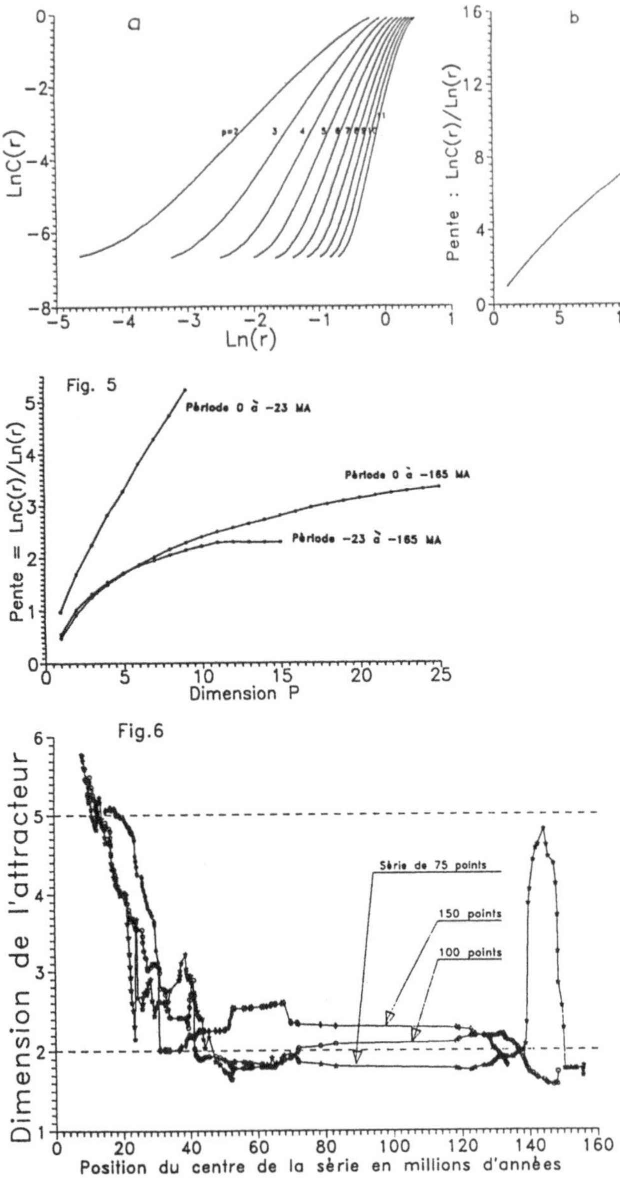

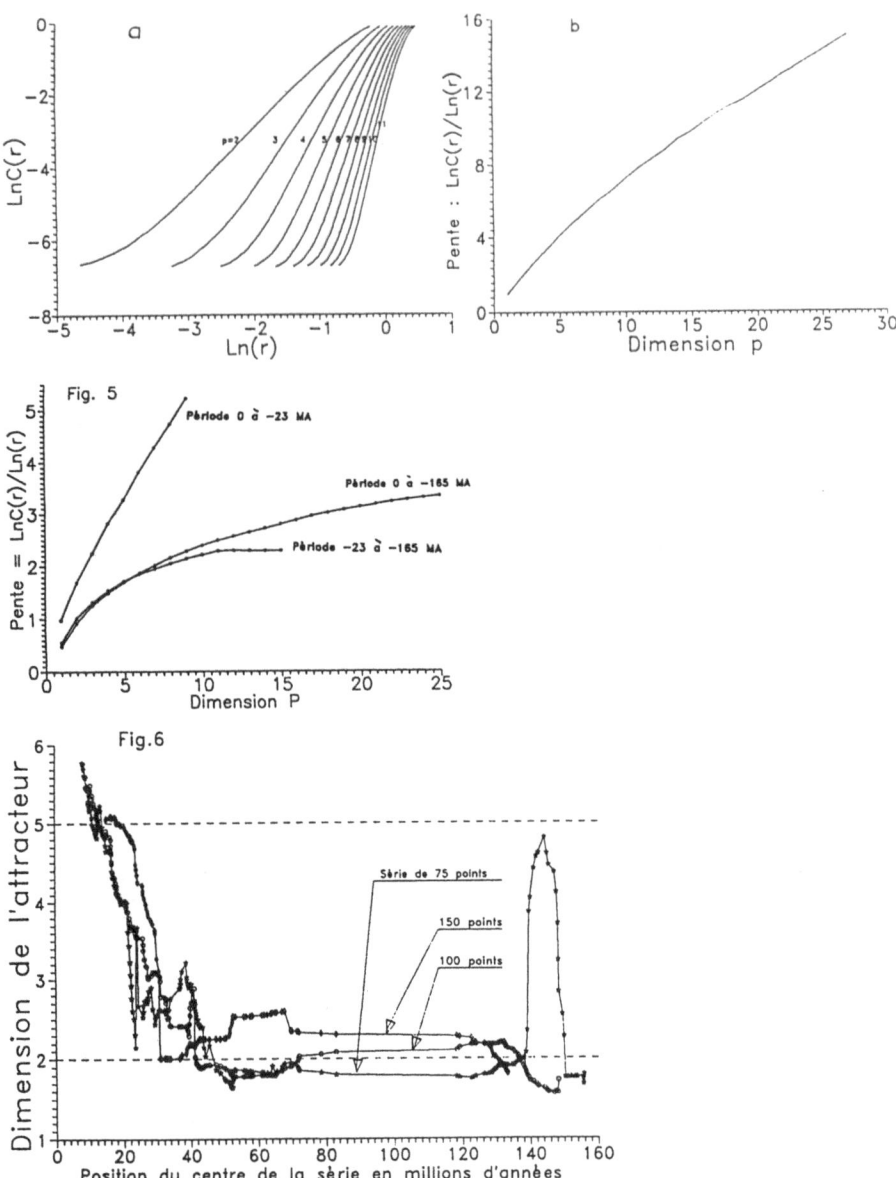

Figure 1.9: *Application of the Grassberger et Procaccia method to the geomagnetic field reversals sequence (Cox's scale).* The correlation fonction slope graphs as fonction of the phase space embedding dimensione are given: a) for the complete data set, b) using a sliding window technique (Dubois et Pambrun, 1990).

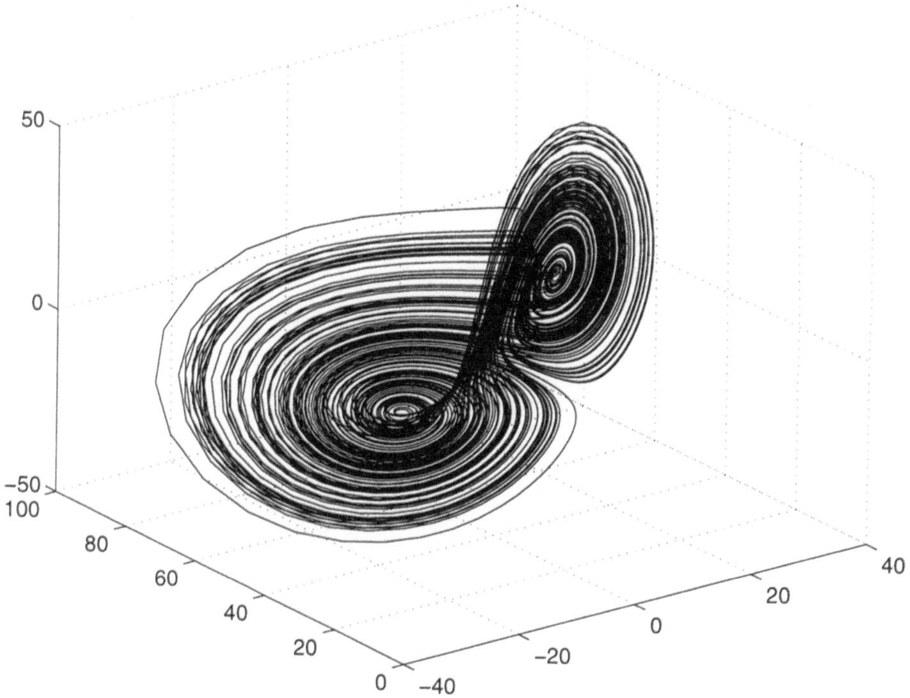


The Lorenz Attractor – The parameters are the same as in page 4, but the perspective representation is different. It shows in the phase space shows 10,000 points discrete phase trajectories (after David AUBERT, 1997).

# Part I

# Foundations

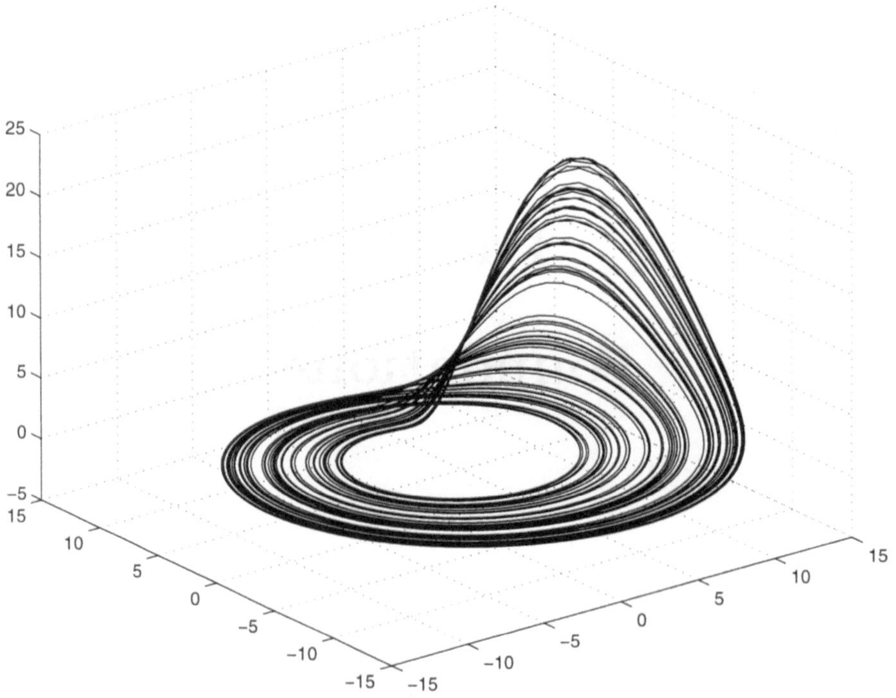


The Rössler Attractor – The parameters (see the text) are: $a = 0.385$ ; $b = 2$ ; $c = 4$. The perspective representation in the phase space shows 10,000 points discrete phase trajectories (after David AUBERT, 1997).

# Chapter 2

# Basic Mathematical Facts

## 2.1 Elements of Set Theory

### 2.1.1 Sets

The important role which the notion of a set plays in modern mathematics and artificial intelligence theory is defined not only by the fact that set theory itself became a wide and self-organized system, but also by the influence which set theory has on the mathematics in general.
Here we just give very basic definitions and other elements of set theory which will be used in this book. For more information on this matter the reader may refer to [117].
The notion of a set is so general, that there is no way to define it using terms that are not simply synonyms of the word *set*. Therefore, we mean that the notion of a set is intuitively clear. We symbolize *sets* by capital letters $\mathcal{A}$, $\mathcal{B}$, $\mathcal{C}$ $\cdots$ and their *elements* by small letters $a$, $b$, $c$, $d$, $\cdots$.
We write $a \in \mathcal{A}$ or $\mathcal{A} \ni a$ for the fact that the element $a$ belongs to the set $\mathcal{A}$. The record $a \notin \mathcal{A}$, $\mathcal{A} \not\ni a$ (or $a\bar{\in}\mathcal{A}$, $\mathcal{A}\bar{\ni}a$) means that the element $a$ does not belong to the set $\mathcal{A}$. If all the elements $a \in \mathcal{A}$ also belong to $\mathcal{B}$ (*e.g* $a \in \mathcal{B}$) we say that $\mathcal{A}$ is a *subset* of $\mathcal{B}$ and write $\mathcal{A} \subset \mathcal{B}$ (the case $\mathcal{A} = \mathcal{B}$ is not excluded). For example the set of integers is a subset of the set of all real numbers.
Sometimes we do not know whether a set contains at least one element or not (for example a set of roots of an equation). Therefore, we introduce the notion of *empty set*, which does not contain any element. We symbolize it by $\emptyset$. Any set has $\emptyset$ as its subset.
Let $\mathcal{A}$ and $\mathcal{B}$ be arbitrary sets. By their *union* $\mathcal{C} = \mathcal{A} \bigcup \mathcal{B}$ we call the set of all elements that belong to at least one of the sets $\mathcal{A}$ or $\mathcal{B}$ (see figure 2.1).

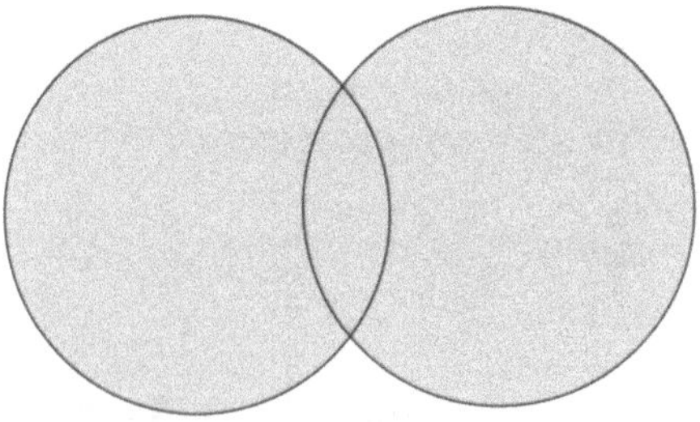

Figure 2.1: *Union of sets.* $\mathcal{C} = \mathcal{A} \bigcup \mathcal{B}$.

In the same way we define the union of any (finite or infinite) number of sets : if $\mathcal{A}_\alpha$ are arbitrary sets, then their union $\bigcup_\alpha \mathcal{A}_\alpha$ is the set of the elements, each of which belongs to at least one of the sets $\mathcal{A}_\alpha$.
We call by *intersection* $\mathcal{C} = \mathcal{A} \bigcap \mathcal{B}$ of the sets $\mathcal{A}$ and $\mathcal{B}$ the set of all the elements which belong to $\mathcal{A}$ and $\mathcal{B}$ simultaneously (see figure 2.2).

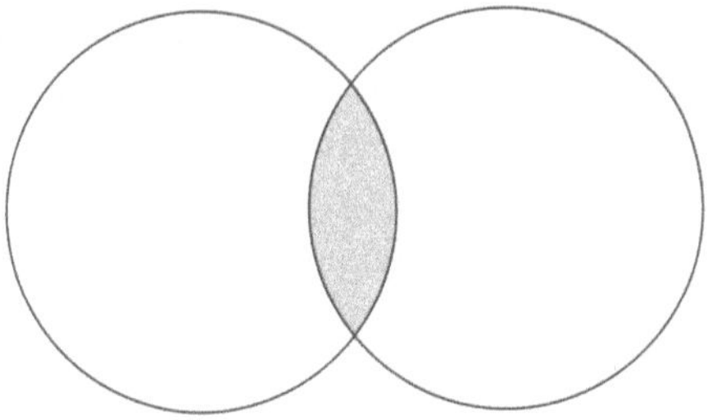

Figure 2.2: *Intersection of sets.* $\mathcal{C} = \mathcal{A} \bigcap \mathcal{B}$.

By the intersection of any (finite or infinite) number of sets $\mathcal{A}_\alpha$ we call the set $\bigcap_\alpha \mathcal{A}_\alpha$ of the elements that belong to each of the sets $\mathcal{A}_\alpha$.
The operations of union and intersection of sets are commutative and associative by definition *e.g.*

$$\mathcal{A} \bigcup \mathcal{B} = \mathcal{B} \bigcup \mathcal{A}; \quad (\mathcal{A} \bigcup \mathcal{B}) \bigcup \mathcal{C} = \mathcal{A} \bigcup (\mathcal{B} \bigcup \mathcal{C}). \tag{2.1}$$

$$\mathcal{A} \bigcap \mathcal{B} = \mathcal{B} \bigcap \mathcal{A}; \quad (\mathcal{A} \bigcap \mathcal{B}) \bigcap \mathcal{C} = \mathcal{A} \bigcap (\mathcal{B} \bigcap \mathcal{C}). \tag{2.2}$$

In addition they are distributive

$$(\mathcal{A}\bigcup\mathcal{B})\bigcap\mathcal{C}=(\mathcal{A}\bigcap\mathcal{C})\bigcup(\mathcal{B}\bigcap\mathcal{C})\,. \tag{2.3}$$

$$(\mathcal{A}\bigcap\mathcal{B})\bigcup\mathcal{C}=(\mathcal{A}\bigcup\mathcal{C})\bigcap(\mathcal{B}\bigcup\mathcal{C})\,. \tag{2.4}$$

The equality $\mathcal{A}=\mathcal{B}$ means that $\mathcal{A}\subset\mathcal{B}$ and $\mathcal{B}\subset\mathcal{A}$.
Let us prove for example the equality (2.3) Suppose that an element $x$ belongs to the set which stays in the left part of (2.3) *e.g.* $x\in(\mathcal{A}\bigcup\mathcal{B})\bigcap\mathcal{C}$. Therefore $x$ belongs to $\mathcal{C}$ and, in addition, at least to one of the sets $\mathcal{A}\bigcap\mathcal{C}$ or $\mathcal{B}\bigcap\mathcal{C}$. This exactly means that $x$ is included into the right part of the equality (2.3).
On the other hand let $x\in(\mathcal{A}\bigcap\mathcal{C})\bigcup(\mathcal{B}\bigcap\mathcal{C})$. Then $x\in\mathcal{A}\bigcap\mathcal{C}$ or $x\in\mathcal{B}\bigcap\mathcal{C}$. Therefore $x\in\mathcal{C}$ and, in addition, $x$ belongs either to $\mathcal{A}$ or to $\mathcal{B}$, *e.g.* $x\in\mathcal{A}\bigcup\mathcal{B}$. Therefore $x\in(\mathcal{A}\bigcup\mathcal{B})\bigcap\mathcal{C}$ and equality (2.3) is proved. In a similar way, it is simple to verify the equality (2.4)

Another operation in the set theory is *substraction.* We define the *difference* $\mathcal{C}=\mathcal{A}\setminus\mathcal{B}$ of the sets $\mathcal{A}$ and $\mathcal{B}$ to be the set of those elements from $\mathcal{A}$, which do not belong to $\mathcal{B}$ (see figure 2.3). Here, generally speaking, we do not assume that $\mathcal{A}\supset\mathcal{B}$. Sometimes the symbolization $\mathcal{A}-\mathcal{B}$ is used for the substraction.

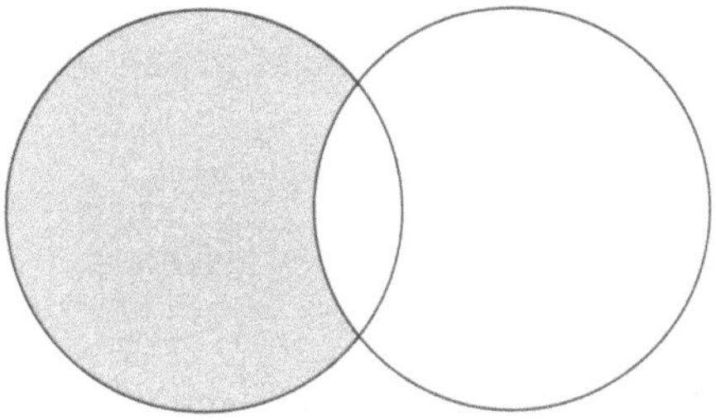

Figure 2.3: *Difference of sets.* $\mathcal{C}=\mathcal{A}\setminus\mathcal{B}$.

Here, generally speaking, we do not assume that $\mathcal{A}\supset\mathcal{B}$. Instead of $\mathcal{A}\setminus\mathcal{B}$ sometimes the symbolization $\mathcal{A}-\mathcal{B}$ is used.
The fourth operation of the set theory is the *symmetrical difference* of sets $\mathcal{A}$ and $\mathcal{B}$. It is defined as the union of the differences $\mathcal{A}\setminus\mathcal{B}$ and $\mathcal{B}\setminus\mathcal{A}$ (see figure 2.4).

$$\mathcal{A}\Delta\mathcal{B}=(\mathcal{A}\setminus\mathcal{B})\bigcup(\mathcal{B}\setminus\mathcal{A})\,. \tag{2.5}$$

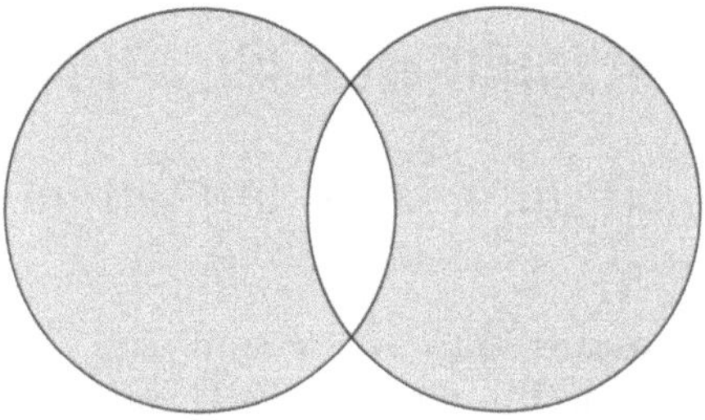

Figure 2.4: *Symmetrical Difference of sets.* $\mathcal{C} = \mathcal{C} = \mathcal{A}\Delta\mathcal{B}$.

It is not complicated to show that the equality (2.5) may be rewritten in the appearance

$$\mathcal{A}\Delta\mathcal{B} = (\mathcal{A}\bigcup\mathcal{B}) \setminus (\mathcal{A}\bigcap\mathcal{B})$$

Often, we consider the sets which are the subsets of a bigger set $\mathcal{S}$. One example is the sets of the points on a line. In this case the difference $\mathcal{S} \setminus \mathcal{A}$ is called the *supplement* (or *closure*) of the set $\mathcal{A}$ and is denoted by $C\mathcal{A}$, $\mathcal{A}'$, or $\bar{\mathcal{A}}$.
The following duality principle plays an important role in set theory and its applications:

1. The supplement to a union is equal to the corresponding intersection of the supplements

$$\mathcal{S} \setminus \bigcup_{\alpha} \mathcal{A}_{\alpha} = \bigcap_{\alpha} (\mathcal{S} \setminus \mathcal{A}_{\alpha}) \tag{2.6}$$

2. The supplement to an intersection is equal to the union of the corresponding supplements.

$$\mathcal{S} \setminus \bigcap_{\alpha} \mathcal{A}_{\alpha} = \bigcup_{\alpha} (\mathcal{S} \setminus \mathcal{A}_{\alpha}) \tag{2.7}$$

From any theorem concerning a system of subsets of a fixed set $\mathcal{S}$ we obtain another dual theorem using the duality principle. The dual theorem we obtain by changing the sets according to the formulas (2.6) and (2.7). In other words we substitute all the sets by their supplements, unions by their intersections, intersections by the corresponding unions.

Let us prove the formula (2.6). Suppose $x \in \mathcal{S} \setminus \bigcup_\alpha \mathcal{A}_\alpha$. This means that $x \notin \bigcup_\alpha \mathcal{A}_\alpha$, in other words $x \notin \mathcal{A}_\alpha$ for any $\alpha$. Therefore $x \in \mathcal{S} \setminus \mathcal{A}_\alpha$ for any $\alpha$ and $x \in \bigcap_\alpha (\mathcal{S} \setminus \mathcal{A}_\alpha)$. On the other hand, suppose $x \in \bigcap_\alpha (\mathcal{S} \setminus \mathcal{A}_\alpha)$. That means that $x \in \mathcal{S} \setminus \mathcal{A}_\alpha$ for any $\alpha$. Therefore, $x$ does not belong to any of the sets $\mathcal{A}_\alpha$, *e.g.* $x \notin \bigcup_\alpha \mathcal{A}_\alpha$. Therefore $x \in \mathcal{S} \setminus \bigcup_\alpha \mathcal{A}_\alpha$ and we have proven the equality (2.6). The equality (2.7) can be proven in the exactly same way.

## 2.1.2 Functions

In mathematical analysis the notion of function is introduced as follows. Let $\mathcal{X}$ be a subset in the set of real numbers. We say that the function $f$ is defined on $x$ if to any number $x \in \mathcal{X}$ there corresponds a unique number $y = f(x)$. The set $\mathcal{X}$ is called the domain of definition of $f$, and the set $\mathcal{Y}$ is the domain of values of $f$.

If instead of the set of real numbers, we consider sets of any nature, then we obtain the most general notion of function. Let $\mathcal{M}$ and $\mathcal{N}$ be two arbitrary sets. We say that on $\mathcal{M}$ the function $f$ with the values from $\mathcal{N}$ is defined, if to every $x \in \mathcal{M}$ there corresponds one and only one element $y \in \mathcal{N}$. For sets of arbitrary nature, the terms map or mapping are often used instead of the term function. Specifying the nature of the sets $\mathcal{M}$ and $\mathcal{N}$ we introduce particular types of functions such as vector-functions, measures, operators etc.

For a function (or mapping) from $\mathcal{M}$ to $\mathcal{N}$ the symbolization

$$f \, : \, \mathcal{M} \to \mathcal{N}$$

is used.

If $a \in \mathcal{M}$, then the corresponding element $b = f(a) \in \mathcal{N}$ is called the *image* of $a$. The set of all $a \in \mathcal{M}$ such that $f(a) = b$ is called the *co-image* of $b$ and is symbolized as $f^{-1}(b)$.

In general we symbolize by $\{x \, : \, \Omega\}$ the set of elements $x$ that satisfy a condition $\Omega$. It can also be specified $\{x \in \mathcal{A} \, : \, \Omega\}$, where $\mathcal{A}$ is a particular set. In this way, we have

$$f^{-1}(b) = \{a \in \mathcal{M} \, : \, f(a) = b\}.$$

Let $\mathcal{A}$ be a subset from $\mathcal{M}$. The set $\{f(a) \, : \, a \in \mathcal{A}\}$ is called the image of $\mathcal{A}$ and is symbolized by $f(\mathcal{A})$ or *im* $f$. At the same time, for every $\mathcal{B} \subset \mathcal{N}$, we define the co-image of $\mathcal{B}$ as

$$f^{-1}(\mathcal{B}) = \{a \in \mathcal{M} \ : \ f(a) \in \mathcal{B}\}$$

Below we describe just the most general features of mappings.
We use the following terminology. We say that $f$ is a mapping of the set $\mathcal{M}$ "onto" the set $\mathcal{N}$, if $f(\mathcal{M}) = \mathcal{N}$. Such mapping is also called a *surjective mapping.* In general, when $f(\mathcal{M}) \subset \mathcal{N}$ we say that $f$ is a mapping of $\mathcal{M}$ "into" $\mathcal{N}$.
If for any two different elements $x_1$ and $x_2$, where $x_1, x_2 \in \mathcal{M}$ their images are also different *e.g.* if $y_1 = f(x_1)$, $y_2 = f(x_2)$, $y_1 \neq y_2$, then $f$ is called *an injective mapping.*

Throughout the entire book, the following theorems will be useful.

**Theorem 2.1.1** The co-image of the union of two sets is equal to the union of their co-images

$$f^{-1}(\mathcal{A} \bigcup \mathcal{B}) = f^{-1}(\mathcal{A}) \bigcup f^{-1}(\mathcal{B}).$$

*Proof* Let $x \in f^{-1}(\mathcal{A} \bigcup \mathcal{B})$. That means that $f(x) \in \mathcal{A} \bigcup \mathcal{B}$. In other words $f(x) \in \mathcal{A}$ or $f(x) \in \mathcal{B}$. But in this case $x$ belongs at least to one of the sets $f^{-1}(\mathcal{A})$ or $f^{-1}(\mathcal{B})$. Therefore $x \in f^{-1}(\mathcal{A} \bigcup f^{-1}(\mathcal{B})$.
On the other hand if $x \in f^{-1}(\mathcal{A}) \bigcup f^{-1}(\mathcal{B})$ then $x$ belongs at least to one of the sets $f^{-1}(\mathcal{A})$ or $f^{-1}(\mathcal{B})$ *e.g.* $f(x)$ belongs at least to one of the sets $\mathcal{A}$ or $\mathcal{B}$. Therefore $f(x) \in \mathcal{A} \bigcup \mathcal{B}$ and $x \in f^{-1}(\mathcal{A}) \bigcup (\mathcal{B})$.

**Theorem 2.1.2** The co-image of the intersection of two sets is equal to the intersection of their co-images.

$$f^{-1}(\mathcal{A} \bigcap \mathcal{B}) = f^{-1}(\mathcal{A}) \bigcup f^{-1}(\mathcal{B}).$$

*Proof* If $x \in f^{-1}(\mathcal{A}) \bigcap (\mathcal{B})$ then $f(x) \in \mathcal{A} \bigcap \mathcal{B}$ *e.g.* $f(x) \in \mathcal{A}$ and $f(x) \in \mathcal{B}$. Therefore $x \in f^{-1}(\mathcal{A})$ and $x \in f^{-1}(\mathcal{B})$. In other words $x \in f^{-1}(\mathcal{A}) \bigcap f^{-1}(\mathcal{B})$.
On the other hand, if $x \in f^{-1}(\mathcal{A}) \bigcap f^{-1}(\mathcal{B})$ *e.g* $x \in f^{-1}(\mathcal{A})$ and $x \in f^{-1}(\mathcal{B})$. Therefore $f(x) \in \mathcal{A}$ and $f(x) \in \mathcal{B}$ or $f(x) \in \mathcal{A} \bigcap \mathcal{B}$. That means that $x \in f^{-1}(\mathcal{A} \bigcap \mathcal{B})$.
Theorems 2.1.1 and 2.1.2 are true for any number of sets:

$$f^{-1}(\bigcup_{\alpha} \mathcal{A}_\alpha) = \bigcup_{\alpha} f^{-1}(\mathcal{A}_\alpha)$$

$$f^{-1}(\bigcap_{\alpha} \mathcal{A}_\alpha) = \bigcap_{\alpha} f^{-1}(\mathcal{A}_\alpha).$$

**Theorem 2.1.3** The image of the union of two sets is equal to the union of their images

$$f(\mathcal{A} \bigcup \mathcal{B}) = f(\mathcal{A}) \bigcup f(\mathcal{B}).$$

*Proof* If $y \in f(\mathcal{A} \bigcup \mathcal{B})$ then $y = f(x)$, where $x$ belongs to at least one of the sets $\mathcal{A}$ or $\mathcal{B}$. Therefore, $y = f(x) \in f(\mathcal{A}) \bigcup f(\mathcal{B})$. On the other hand, if $y \in f(\mathcal{A}) \bigcup f(\mathcal{B})$ then $y = f(x)$, where $x$ belongs to at least one of the sets $\mathcal{A}$ or $\mathcal{B}$. Therefore $x \in \mathcal{A} \bigcup \mathcal{B}$ and $y = f(x) \in f(\mathcal{A} \bigcup \mathcal{B})$.
At the same time the image of the intersection of two sets, generally speaking, is not equal to the intersection of their images. An example is the projection of the plane to the axis 0x. Then the segments $0 \leq x \leq 1$, $y = 0$ and $0 \leq x \leq 1$, $y = 1$ do not intersect, but their images are equal.
If as usual $f : \mathcal{M} \to \mathcal{N}$ and $\mathcal{A} \subset \mathcal{N}$, then the following important formula holds

$$f^{-1}(\mathcal{N} \setminus \mathcal{A}) = \mathcal{M} \setminus f^{-1}(\mathcal{A})$$

This formula can be proven in exactly the same way as theorems 2.1.1-2.1.3.

## 2.2 Elements of Linear Algebra

The reader can find in [65] the classical theory of linear algebra in all its details.
Here we just give basic definitions and facts from this beautiful theory necessary for understanding the techniques described in this book.

**Definition 2.2.1** $K$–dimensional Euclidean space $\Re^K$ over the field of the real numbers $\Re$ is the set of all ordered sets of $K$ real numbers $\mathcal{Z} = (z_1, \cdots, z_K)$ with three operations (2.8), (2.9), (2.10).
Herein, $\mathcal{Z}$ is called a point or a vector, and the numbers $(z_1, \cdots, z_K)$ are called the coordinates. The null-vector $(0,\ 0 \cdots,\ 0)$ (or the origin of the coordinates) is symbolized by $\mathcal{O}^K$ or sometimes by just 0.

Addtition of vectors $z \in \Re^K$, their multiplication by real numbers, and their scalar product are defined by the following formulas :

$$z' + z'' = (z_1' + z_1'', \cdots, z_K' + z_K''), \tag{2.8}$$

$$\lambda z = (\lambda z_1, \cdots, \lambda z_K), \tag{2.9}$$

$$< z', z'' >= \sum z_i' z_i'', \quad i = 1, \cdots, K, \tag{2.10}$$

where $z' = (z_1', \cdots, z_K')$, $z'' = (z_1'', \cdots, z_K'')$, $z = (z_1, \cdots, z_K) \in \Re^K$, $z,\ z'\ z'' \in \Re^K$.
If we consider only the operations (2.8) and (2.9), then $\Re^K$ is called a linear space.
The vector $z = \sum \lambda_i z^i$, $\lambda_i \in \Re$, $z^i \in \Re^K$ $i = 1, \cdots, K$ is called a linear combination of the vectors $z^1, \cdots, z^K$. A linear combination is degenerative if all the coefficients $\lambda_i$ are equal to zero. In this case $z = 0^K$.
The *linear span* of the vectors $z^1, \cdots, z^r, \in \Re^K$ is symbolized by $lin(z^1, \cdots, z^r)$ and defined as the set of all possible linear combinations of $z^1, \cdots, z^r$.
A linearly independent system of vectors $z^1, \cdots, z^K \in \Re^K$ is defined by one of two equivalent requirements :

(a) only degenerative linear combinations of $z', \cdots, z^K$ are equal to $0^K$.

(b) among the vectors $z^1, \cdots, z^K$ there is at least one which can be represented as a linear combination of the others.

A linearly independent system in $\Re^K$ cannot include more than $K$-vectors.
The basis $\mathcal{B}_\mathcal{S}$ of a subset $\mathcal{S} \subseteq \Re^K$ including $\Re^K$ itself is defined by one of the following three equivalent definitions :

(a) $\mathcal{B}_\mathcal{S}$ is a set of linearly independent vectors through which all $z \in \mathcal{S}$ can be represented as their linear combination.

(b) $\mathcal{B}_\mathcal{S}$ is a minimal set of vectors $z \in \mathcal{S}$ which satisfies the equality $lin(\mathcal{B}_\mathcal{S}) = \mathcal{S}$.

(c) $\mathcal{B}_\mathcal{S}$ is a maximal set of linearly independent vectors from $\mathcal{S}$.

In (b) and (c) maximum and minimum can be understood in the sense of the quantity of vectors as well as in the sense of the relation of inclusion.

All the bases in $\mathcal{S} \subseteq \Re^K$ have the same number $n$ of vectors and this number is not bigger than $K$. The number $\ell$ is called the dimension of $\mathcal{S}$ and is symbolized by $\dim \mathcal{S}$. The dimension of the whole space $\dim \Re^K = K$. The standard basis $\mathcal{B}^K$ in $\Re^K$ consists of $K$ vectors

$$\begin{aligned} \mathcal{B}^K = \{e^i,\ i = 1,\ \cdots,\ K\} &= \{e^i = (e_1,\ \cdots,\ e_K) :\ j, i = 1,\ \cdots,\ K\} \\ e^i_j = \delta^i_j &= \begin{cases} 1, & \text{if } i = j \\ 0, & \text{if } i \neq j \end{cases} \end{aligned} \tag{2.11}$$

$\mathcal{B}^K$ is also an orthogonal basis, $< e^i, e^j >= \delta^i_j$.
In particular $\|e^i\|^2 = 1$.
For every vector $z \in \Re^K$, its coefficients of decomposition in the basis $\mathcal{B}^K$ are equal to the coordinates $(z_1,\ \cdots,\ z_K)$, $z_i =< z,\ e^i >$.
The important notion of a subspace $\mathcal{L} \subset \Re^K$ is introduced by one of the following equivalent definitions :

**Definition 2.2.2 (a)** $\mathcal{L}$ is a linear space by itself *e.g.* $\mathcal{L}$ is closed for the vector operations of addition and multiplication by real numbers.

**Definition 2.2.2 (b)** For every $z', z^2 \in \mathcal{L}$ their linear cover $lin(z', z^2) \subset \mathcal{L}$.

**Definition 2.2.2 (c)** Let $r \geq 2$ be fixed. Then for every $z',\ \cdots,\ z^r) \in \Re^K$, $lin(z',\ \cdots,\ r^K) \subset \Re^K$.

It is clear that intersection of any (may be infinite) number of suspaces is a subspace.
*Linear cover* of an arbitrary set $\mathcal{S} \subset \Re^K$ is symbolized by $lin\mathcal{S}$ and defined by one of the following two equivalent conditions.

(a) $lin\mathcal{S}$ is a minimal subspace $\mathcal{L}$ such as $\mathcal{L} \supset \mathcal{S}$.

$$lin\mathcal{S} = \min\{\mathcal{L} :\ \mathcal{L} \supset \mathcal{S}\}$$

(b) $lin\mathcal{S}$ is the set of all possible linear combinations of the finite subsets $\mathcal{S}' \subset \mathcal{S}$.

It is not complicated to show that the equality $lin\mathcal{S} = \mathcal{S}$ holds if and only if $\mathcal{S}$ is a subspace.

The determinant of a square $(K \times K)$ matrix $\mathbf{A} = \|a_{ij}\|$ with real coefficients $a_{ij}$, $i, j = 1, \cdots, K$ is symbolized by $\det \mathbf{A}$ or $\det \|a_{ij}\|$. It is defined by the formula

$$\det \mathbf{A} = \sum_{\sigma} (-1)^{|\sigma|} \prod_{i} a_{i\sigma(i)} \tag{2.12}$$

where the multiplication is done for all $i \in K$ and the summation for all substitutions $\sigma$ on the set $K$ (by $\mid \sigma \mid$ the number of inversions of the substitution $\sigma$: is symbolized [see [162]].

The following conditions are equivalent

$\det \mathbf{A} = 0$ (2.2.3.)

$K$ vectors in $\Re^K$, which correspond to lines (columns) of the matrix $\mathbf{A}$ create a basis in $\Re^K$ (2.2.4.)

A matrix which satisfies the conditions (2.2.3.) and (2.2.4.) is called *non-degenerative.*

The following interpretation of determinant holds. Module $\mid \det \mathbf{A} \mid$ is equal to the volume of the parallelepiped in $\Re^K$ with a summit in $0^K$ defined by $K$-vector lines (vector-columns) of the matrix $\mathbf{A}$. The sign( $\det \mathbf{A}$) corresponds to right or left orientation of the basis in $\Re^K$ defined by these vectors. If they are not a basis, then $\det \mathbf{A} = 0$.

The rank of a rectangular $(n \times m)$ matrix $\mathbf{A} = \|a_{ij}\|$ which we symbolize by rank $A$ is, in some sense, a generalization of the notion of determinant to the whole set of rectangular matrices. We say that rank $A = r$ if one of the following three equivalent conditions takes place :

(a) There is a non-degenerative sub matrix in $\mathbf{A}$ of the order $r \times r$ (but not more).

(b) There are $r$ linearly independent vector-lines in $\mathbf{A}$ (but not more).

(c) There are $r$ linearly independent vector-columns in $\mathbf{A}$ (but not more).

It is clear that rank $A < \min(m, n)$.

## 2.3 Elements of Boolean Algebra

**Definition 2.3.1** Boolean algebra is a set of elements $\{\alpha, \beta, \gamma, \cdots\}$ with two binary operations ($+$ and $*$), a unitary operation $(d \to \overline{\alpha})$, two special elements 0 and $e$ and the relation $\alpha \subset \beta$. In Boolean algebra the following axioms hold:

$$
\begin{array}{ll}
1a) & \alpha + \beta = \beta + \alpha \\
1b) & (\alpha + \beta) + \gamma = \alpha + (\beta + \gamma) \\
1c) & \alpha + \alpha = \alpha
\end{array}
$$

$$
\begin{array}{ll}
2a) & \alpha * \beta = \beta * \alpha \\
2b) & (\alpha * \beta) * \gamma = \alpha * (\beta * \gamma) \\
2c) & \alpha * \alpha = \alpha
\end{array}
$$

$$
\begin{array}{ll}
3a) & (\alpha + \beta) * \gamma = \alpha * \gamma + \beta * \gamma \\
3b) & (\alpha * \beta) + \gamma = (\alpha + \gamma) * (\beta + \gamma)
\end{array}
$$

$\exists e$ and $\exists 0$ such as $\forall \alpha$ :

$$
\begin{array}{ll}
4a) & \alpha * e = \alpha \quad \alpha + e = e \\
4b) & \alpha + 0 = \alpha \quad \alpha * 0 = 0
\end{array}
$$

$$5a) \quad \overline{\overline{\alpha}} = \alpha \quad 5b) \quad \overline{e} = 0 \quad 5c) \quad \overline{0} = e$$

**Morgan's law** :

$$
\begin{array}{llll}
6a) & \overline{\alpha + \beta} = \overline{\alpha} * \overline{\beta} & 6b) & \overline{\alpha * \beta} = \overline{\alpha} + \overline{\beta} \\
6c) & \alpha + \overline{\alpha} = e & 6d) & \alpha * \overline{\alpha} = 0
\end{array}
$$

$$7a) \quad \alpha \subset \alpha \quad 7b) \quad \alpha \subset \beta,\ \beta \subset \gamma \Rightarrow \alpha \subset \gamma$$

$$7c) \quad \alpha \subset \beta, \quad \beta \subset \alpha \Rightarrow \alpha = \beta$$

$$8a) \quad \alpha + \beta \supset \alpha \qquad 8b) \quad \alpha * \beta \supset \alpha$$

$$8c) \quad \alpha \subset \beta \Rightarrow \alpha + \beta = \beta \text{ and } \alpha * \beta = \alpha$$

$$9) \quad \alpha \supset \beta \quad \Rightarrow \quad \overline{\alpha} \subset \overline{\beta}$$

If, in the Definition 2.3.1, we take into consideration only the conditions 3a and 3b, we obtain a set which is called as **grid**.
An example of a Boolean algebra is the algebra of all the subsets of a given set $e$. Here $+$ is unification, $*$ is intersection, $\alpha \rightarrow \overline{\alpha}$ is consideration of the complementary set, $0 = \emptyset$ is the empty set, and $\alpha \subset \beta$ is the relation of sets inclusion.
Another example of Boolean algebra is the algebra of sentences [181]. Here $+$ is disjunction $\bigvee$ ("or"), $*$ is conjunction. $\bigwedge$ ("and"), $\alpha \rightarrow \overline{\alpha}$ is the opposite sentence ("not $\alpha$"), $e$ is the sentence always true, $0$ is the sentence always false and $\subset$ is implication.
The axioms 2-9 in the definition 2.3.1 are not independent. Thus, it is possible to give a shorter, but less visualisable definition of Boolean algebra.

**Definition 2.3.2** Boolean algebra is a set of elements $\{\alpha,\ \beta,\ \gamma,\ \delta,\ \cdots\}$ with two binary operations ($+$ and $*$), one unitary operation ($\alpha \rightarrow \overline{\alpha}$), two special elements ($0$ and $e$) for which the following axioms hold:

$$
\begin{array}{llll}
1) & \alpha+\beta=\beta+\alpha & 2) & \alpha*\beta=\beta*\alpha \\
3) & \alpha*(\beta+\gamma)=\alpha*\beta+\alpha*\gamma & 4) & \alpha+\beta*\gamma=(\alpha+\beta)*(\alpha+\gamma) \\
5) & \alpha+0=\alpha & 6) & \alpha*e=\alpha \\
7) & \alpha+\overline{\alpha}=e & 8) & \alpha*\overline{\alpha}=0.
\end{array}
$$

The axioms in the definition 2.3.2 are independent. The relation $\alpha \subset \beta$ can be introduced using the equations $\alpha+\beta=\beta$ and $\alpha*\beta=\alpha$.

**Definition 2.3.3** Non-complete Boolean algebra is a set $\{\alpha,\ \beta,\ \gamma,\ \cdots\}$ with all conditions listed in Definition 1.3.1, except the conditions 6c and 6d.

Here are two examples of non-complete Boolean algebras.

1. Let $N$ be a fixed integer, and $\{\alpha,\ \beta,\ \gamma,\ \cdots\}$ be the set of all divisors of $N$. We introduce the following operations : $\alpha+\beta = \mathcal{SCM}(\alpha,\beta)$ (Smallest Common Multiplier of $\alpha$ and $\beta$) and $\alpha*\beta = \mathcal{LCD}$ (Largest Common Divisor of $\alpha$ and $\beta$). We denote $e = N,\ 0 = 1,\ \overline{\alpha} = N/\alpha,\ \alpha \subset \beta \Leftrightarrow \beta$ is a divisor of $\alpha$.

2. We consider the set of all $x \in [0,1]$ and symbolize $\alpha+\beta = \max\{\alpha,\ \beta\}$, $\alpha*\beta = \min\{\alpha,\ \beta\}$, $1=1,\ 0=0$ and $\overline{\alpha} = 1-\alpha,\ \alpha \subset \beta \Leftrightarrow \alpha \subseteq \beta$ as a subset.

**Definition 2.3.4** A function $\mathcal{F} = \mathcal{F}(X_1, \cdots, X_n)$ with values 0 or 1 and the arguments $X_1, \cdots, X_n$ which are also equal to 0 or 1 is called a Boolean function.
A natüral way to define a Boolean function is to introduce a truth table. Here is an example of such a table:

| $x_1$ | $x_2$ | $x_3$ | $\mathcal{F}(x_1, x_2, x_3)$ |
|---|---|---|---|
| 0 | 0 | 0 | 0 |
| 0 | 0 | 1 | 1 |
| 0 | 1 | 0 | 0 |
| 0 | 1 | 1 | 1 |
| 1 | 0 | 0 | 0 |
| 1 | 0 | 1 | 0 |
| 1 | 1 | 0 | 1 |
| 1 | 1 | 1 | 1 |

In the set of Boolean functions, we introduce operations and relations as follows:

1. **Disjunction.** $\mathcal{F} = \mathcal{F}_1 \bigvee \mathcal{F}_2$. Here $\mathcal{F} = 1$ if and only if $\mathcal{F}_1 = 1$ or $\mathcal{F}_2 = 1$. In other words $1 \bigvee 0 = 1$ ; $0 \bigvee 1 = 1$ ; $1 \bigvee 1 = 1$ ; $0 \bigvee 0 = 0$.

2. **Conjunction.** $\mathcal{F} = \mathcal{F}_1 \bigwedge \mathcal{F}_2$. Here $\mathcal{F} = 1$ if and only if $\mathcal{F}_1 = 1$ and $\mathcal{F}_2 = 1$. In other words $1 \bigwedge 0 = 0$ ; $0 \bigwedge 1 = 0$ ; $1 \bigwedge 1 = 1$ ; $0 \bigwedge 0 = 0$.

3. **Rejection.** Here $\overline{\mathcal{F}} = 1$ if and only if $\mathcal{F} = 0$. In other words $\overline{0} = 1$ ; $\overline{1} = 0$.

4. **Implication.** Here $\mathcal{F}_1 \Rightarrow \mathcal{F}_2$ if and only if $\mathcal{F}_1 = 1 \Rightarrow \mathcal{F}_2 = 1$.

We also introduce the unit Boolean function $\mathcal{E}$ that is equal to 1 for every argument and the null function that is equal to 0 for every argument.

**Theorem 2.3.1** The set of all Boolean functions of $n$ arguments is a Boolean algebra with operations disjunction $(+)$, conjunction $(*)$, and rejection $(\mathcal{F} \rightarrow \overline{\mathcal{F}})$ the relation of implication.
The proof of the theorem is actually the verification of the axioms of the definition 2.3.2 which is given in [181].

The Boolean functions $\mathcal{F}(x) = x$ and $\mathcal{F}(x) = \overline{x}$ are called "elementary". By elementary disjunction (elementary conjunction) we mean disjunction (conjunction) of any finite number of elementary functions. Thus, for example, $x_1 \bigvee x_2 \bigvee \overline{x_3} \bigvee \overline{x_4}$ is elementary disjunction, and $x_1 \bigwedge x_2 \bigwedge \overline{x_3} \bigwedge \overline{x_4}$ is an elementary conjunction.
Sometimes we do not write the symbol of conjunction considering that $x \bigwedge y \doteq xy$ by definition.

**Definition 2.3.5** Disjunctive (conjunctive) normal form $\mathcal{DNF}$ ($\mathcal{CNF}$) is a disjunction of a finite number of elementary conjunctions (correspondingly, a conjunction of a finite number of elementary disjunctions).
Examples are the following :

$$\begin{array}{ll} \mathcal{DNF} : & x_1x_2 \bigvee x_1\overline{x_3} \bigvee \overline{x_2}x_3 \\ \mathcal{CNF} : & (x_1 \bigvee x_2 \bigvee x_3)(x_1 \bigvee \overline{x_3})(\overline{x_2} \bigvee x_3). \end{array}$$

**Theorem 2.3.2** Every Boolean function can be defined by $\mathcal{DNF}$ and $\mathcal{CNF}$. Correspondance between a Boolean function and a $\mathcal{DNF}$ is defined by the truth table. Corresponding $\mathcal{CNF}$ is constructed using the axiom of distributivity [see Definition 2.3.1., 3a)-3b)].

**Definition 2.3.6** A Boolean function is called monotonous, if it can be defined by a $\mathcal{DNF}$ without rejections.
An example of a monotonous function is the following :

$$\mathcal{F}(x_1, x_2, x_3, x_4) = x_1x_2x_3 \bigvee x_3x_4 \bigvee x_1x_2x_4 . \tag{2.13}$$

The function $\mathcal{F}^\star = \overline{\mathcal{F}}(\overline{x_1}, \cdots, \overline{x_n})$ is called the dual function to $\mathcal{F}$ where $\mathcal{F} = \mathcal{F}(x_1, \cdots, x_n)$.

**Theorem 2.3.3** If $\mathcal{F}$ is monotonous, then $\mathcal{F}^\star$ is also monotonous.
*Scheme of proof*: We represent $\mathcal{F}$ by $\mathcal{DNF}$ without rejections and apply Morgan's law (see Definition 2.3.1).
Generally, the following conditions hold:
$\mathcal{E}^\star = \mathcal{O}$, $\mathcal{O}^\star = \mathcal{E}$, $\mathcal{F}^{\star\star} = \mathcal{F}$, and the functions $\mathcal{E}$ and $\mathcal{O}$ are monotonous.
At the same time $\mathcal{F} \bigvee \mathcal{F}^\star = \mathcal{E}$, $\mathcal{F} \bigwedge \mathcal{F}^\star = 0$.
Furthermore, the following theorems hold:

**Theorem 2.3.4** The set of monotonous Boolean functions is a non-complete Boolean algebra with operations of disjunction, conjunction, duality and relation of implication.

**Theorem 2.3.5** Let $\alpha_1, \cdots, \alpha_k$ ; $\beta_1, \cdots, \beta_\ell$ be elementary conjunctions and $\mathcal{F} = \alpha_1 \bigvee \cdots \bigvee \alpha_k$, $\mathcal{F}^* = \beta_1 \bigvee \cdots \bigvee \beta_\ell$ – $\mathcal{DNF}$ pairs of dual monotonous functions. Then $\forall i = 1, \cdots, k$, $j = 1, \cdots, \ell$, $\alpha_i$ and $\beta_j$ have common argument.

## 2.4 Metric spaces

Metric space is one of the most important notions of modern mathematics. The generalization of metric space theory leads to the notion of topological space– the basic object of the mathematical discipline called topology. Here, as everywhere in this introductory part of the book, we give only very basic facts on metric spaces and we refer to the book [117] if the reader would like to have more information on the subject.

**Definition 2.4.1** By metric space we call a pair $(\mathcal{X}, \rho)$ where $\mathcal{X}$ is a set (space) and $\rho$ is a metric *e.g.* non-negative real-valued functions $\rho: \mathcal{X} \to \Re$ which satisfies $\forall x, y, z \in \mathcal{X}$ to the following three axioms:

1. $\rho(x,y) = 0$ when and only when $x = y$,
2. $\rho(x,y) = \rho(y,x)$ (axiom of symmetry),
3. $\rho(x,z) = \rho(x,y) + \rho(y,z)$ (axiom of triangle).

We will usually symbolize a metric space $(\mathcal{X}, \rho)$ by $\mathcal{R} = (\mathcal{X}, \rho)$ or just $\mathcal{X}$. Herein are several well known and widely used examples of metric spaces. The reader will find some others in part II of this book.

1. We define for an arbitrary set the metrics

$$\rho(x,y) = \begin{cases} 0, & \text{if } x = y \\ 1, & \text{if } x \neq y \end{cases} \tag{2.14}$$

   This metric space is called *the space of isolated points.*

2. The set of real number $\Re$ with the distance

$$\rho(x,y) = \mid x - y \mid \tag{2.15}$$

   is the metric space symbolized by $\Re^1$.

3. The space $\Re^K$ (see section 2.2) is a metric space with the metrics

$$\rho(x,y) = \sqrt{\sum_{i=1}^{K}(y_i - x_i)^2}\,. \tag{2.16}$$

4. Considering the same set $\Re^K$ and defining the metric by the formula

$$\rho_1(x,y) = \sum_{i=1}^{K} \mid x_K - y_K \mid \tag{2.17}$$

we get another metric space. It is denoted by $\Re_1^K$.

5. Considering again the set $\Re^K$, but with the metric

$$\rho_0(x,y) = \max_{1 \le K \le n} \mid y_K - x_K \mid, \tag{2.18}$$

we obtain the metric space $\Re_0^K$.

6. The set $\mathcal{C}[a,b]$ of all real continuous functions, defined on the segment $[a,b]$ with the metric

$$\rho(f,g) = \max_{a \le t \le b} \mid g(t) - f(t) \mid \tag{2.19}$$

is another type of metric space which is called *a functional space.*

There are many other important examples of different kinds of metric spaces (see []).

Let $\mathcal{X}$ and $\mathcal{Y}$ be metric spaces and $f$ a mapping $f: \mathcal{X} \to \mathcal{Y}$. This mapping is called continuous at a point $x_0 \in \mathcal{X}$, if for every $\epsilon > 0$ there is a $\delta > 0$ such as $\forall x \in \mathcal{X}$ that satisfies the conditions

$$\rho(x, x_0) < \delta$$

and

$$\rho_1(f(x), f(x_0)) < \epsilon$$

(here $\rho$ is the metric in $\mathcal{X}$, and $\rho_1$ is the metric in $\mathcal{Y}$). If a mapping is continuous at all points $x \in \mathcal{X}$ then it is said that $f$ is continuous on $\mathcal{X}$. In the same way we may define a continuous mapping $f(x_1, \cdots, x_n): \mathcal{X}_1 \times \cdots \times \mathcal{X}_n \to \mathcal{Y}$, where $\mathcal{X}_1, \cdots, \mathcal{X}_n, \mathcal{Y}$ are metric spaces.

Let us consider, for example, the function $f(x,y) = \rho(x,y): \mathcal{X} \times \mathcal{X} \to \Re$. This function is continuous which immediately follows from the ineguality

$$| \rho(x,y) - \rho(x_0,y_0) | \leq \rho(x_0,x) + \rho(y_0,y) . \tag{2.20}$$

The ineguality follows from the triangle axiom.
If a mapping $f: \mathcal{X} \to \mathcal{Y}$ is such that it is surjective and injective (see 1.1) then the reverse mapping $\mathcal{X} = f^{-1}(\mathcal{Y})$ exists and $f^{-1}: \mathcal{Y} \to \mathcal{X}$ is also surjective and injective.
If both mappings $f$ and $f^{-1}$ are continuous then $f$ is called *homeomorphic* or a *homeomorphic mapping.* The spaces $\mathcal{X}$ and $\mathcal{Y}$ in such cases are called *homeomorphic spaces.*
An example of a pair of homeomorphic metric spaces is the whole straight line $(-\infty,+\infty)$ and an interval, for example $(-1,+1)$. In this case the homeomorphism is established by the mapping

$$y = \frac{2}{\pi} \arctan x$$

In this book we will deal mostly with the finite metric spaces (*e.g* where the set $\mathcal{X}$ in the definition 1.4.1 is finite, $| \mathcal{X} | < \infty$). It is obvious that all such metric spaces are homeomorphic to each other.
An important particular case of homeomorphic mappings is so called *isometric mappings.*

**Definition 2.4.2** Let $\mathcal{R} = (\mathcal{X},\rho)$, $\mathcal{R}' = (\mathcal{Y},\rho')$ and $f$ be a surjective and injective mapping $f: \mathcal{X} \to \mathcal{Y}$. Then $f$ is called an isometric mapping or *isometry* if

$$\rho(x_1,x_2) = \rho'(f(x_1), f(x_2)) \tag{2.21}$$

for any $x_1, x_2 \in \Re$.
The metric spaces between which it is possible to establish an isometry are called isometric spaces.
It is clear that the finite metric spaces that we will consider in this book are isometric to each other. Furthermore isometry of spaces $\mathcal{R}$ an $\mathcal{R}'$ means that metric connections between their elements are identical.
Only the nature of elements can differ, which is not important as far as theory of metric spaces is concerned. Therefore, all isometric spaces (finite metric spaces in particular) we consider as equal.

**Definition 2.4.3** As an open ball $\mathcal{B}(x_0, r)$ in a metric space $\Re$ we call the set

$$\mathcal{B}(x_0, r) = \{x \in \Re: \rho(x, x_0) < r\} \ . \tag{2.22}$$

The point $x_0$ is called the center of the ball and the number $r$ is called its radius.
A closed ball $\mathcal{B}[x_0, r]$ with the center $x_0$ and radius $r$ is the set

$$\mathcal{B}[x_0, r] = \{x \in \Re: \rho(x, x_0) \leq r\} \ . \tag{2.23}$$

In this book we will deal with finite metric spaces. Therefore, all topological constructions below are defined in metric space of isolated points which simplify the situation. We refer to [117] for the reader interested to learn the metric space theory in the continuous case.

# Part II

# Dynamic Pattern Recognition Problems and Control over Classification Reliability

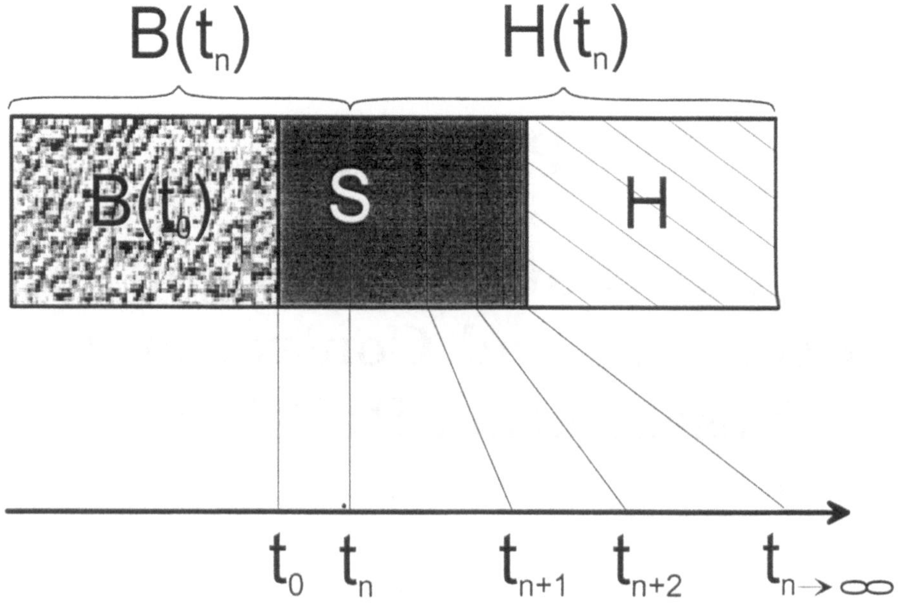


Dynamic pattern recognition problem - $t-$time scale, $\mathcal{W} = \mathcal{W}_0 = \mathcal{B}(t_0) \bigsqcup [\mathcal{S} \bigcup \mathcal{H}]$- initial decomposition (learning material), $\mathcal{W} = \mathcal{B}(t_n) \bigsqcup \mathcal{H}(t_n)$ - current decomposition, $\mathcal{W} = \lim_{t_n \to \infty}[\mathcal{B}(t_n) \bigsqcup \mathcal{H}(t_n)] = [\mathcal{B}(t_0) \bigcup \mathcal{S}] \bigsqcup \mathcal{H} = \mathcal{B} \bigsqcup \mathcal{H}$- final classification (prediction).

# Chapter 3

# “Voting by a Set of Features” Algorithms

## 3.1 Basic Definitions

We shall use below the following symbols and definitions [84, 116]:

- $\{\alpha, \beta, \gamma, \ldots\}$ — the set which consists of the elements $\alpha, \beta, \gamma, \ldots$
- $\emptyset$ — empty set.
- $\mathcal{A} \bigsqcup \mathcal{B}$ — disjunctive unification of the sets $\mathcal{A}$ and $\mathcal{B}$; $\mathcal{A} \bigsqcup \mathcal{B} = \mathcal{A} \bigcup \mathcal{B}$, if $\mathcal{A} \bigcap \mathcal{B} = \emptyset$.
- $\mathcal{A} \Leftrightarrow \mathcal{B}$ —“a statement $\mathcal{A}$ takes place if and only if a statement $\mathcal{B}$ takes place”.
- $\overline{\mathcal{A}} = \mathcal{X} \setminus \mathcal{A}$, where $\mathcal{X}$ is a fixed set.
- $\mid \mathcal{X} \mid$ — number of elements in the finite set $\mathcal{X}$.
- $\mathcal{A} \times \mathcal{B}$ — direct product of the sets $\mathcal{A}$ and $\mathcal{B}$. This is the set of all pairs $\{(a,\ b)\ :\ a \in \mathcal{A},\ b \in \mathcal{B}\}$.
- $\mathcal{P}\{A\}$ — probability of an event $A$.
- $\mathcal{P}\{A \mid B\}$ — probability of an event $A$ in the condition of an event $B$.
- $\Re^n$ — real coordinate $n$-dimensional metric space.

- $\Omega_n$ — set of all vectors $\omega$ which components have the values **0** or **1**. Vectors $\omega$ we call binary. $\omega = (\omega(1),\ \omega(2),\ \dots,\ \omega(n))$, $\omega(i) = 0$ or $1$, $i = 1,\ 2,\ \dots,\ n$.

- $(\mathcal{X},\ \rho)$ —metric space which consists of a set $\mathcal{X}$ and a metric $\rho$.

- $\mathcal{W}$ — set of objects of recognition. The algorithms introduced below are applied for classification of this set.

- $\mathcal{W}_0 \subset \mathcal{W}$ — fixed set which we call the learning material or learning set. We assume that a decomposition $P : \mathcal{W}_O = \mathcal{B}_0 \bigsqcup \mathcal{H}_0$ is given over the learning set. We call it the initial decomposition. We also represent the initial decomposition as the pair $\mathcal{O}_0 = (\mathcal{B}_0,\ \mathcal{H}_0)$.

- $\Omega = \{x_1(w),\ \dots,\ x_N(w)\}$ — parametrization of objects $w \in \mathcal{W}$ by the given functions $X_i(w)$, $i = 1,\ \dots,\ N$, with real or binary values.

- $\Phi : \mathcal{W} \to \Omega_n$ — given mapping which is naturally called "coding of the objects".

  If $\mathcal{A} \subset \mathcal{W}$, then $\Phi(\mathcal{A}) = \{\Phi(w) : w \in \mathcal{A}\}$,

  If $\mathcal{B} \subset \Omega_n$, then $\Phi^{-1}(\mathcal{B}) = \{w \in \mathcal{W} : \Phi(w) \in \mathcal{B}\}$.

The algorithms which we construct below represent the set $\Omega_n$ as a disjunctive unification of sets $\Lambda_\mathcal{B}$ and $\Lambda_\mathcal{H}$, based on analysis of information about the coding $\Phi$ and the sets $\mathcal{B}_0$ and $\mathcal{H}_0$. This analysis is done in a way that characterizes an algorithm. Such decomposition of the set $\Omega_n$ we call the classification obtained by a given algorithm.
The algorithms under consideration depend on the parameters. Values of these parameters have to be fixed before applying the algorithm to the classification of the set $\Omega_n$. Such fixed parameters of the algorithm are called "free" and are symbolized as $\mathcal{C}$. By changing the free parameters for a given algorithm, we obtain, generally speaking, different classifications.
Classifications obtained by an algorithm $\mathcal{A}$ will be denoted as

$$\mathcal{A}(\mathcal{P},\mathcal{C}) = \mathcal{A}(\mathcal{O},\ \mathcal{C})\ :\ \Omega_n = \Lambda_\mathcal{B} \bigsqcup \Lambda_\mathcal{H}\,, \tag{3.1}$$

or as

$$\mathcal{A}(\mathcal{P},\mathcal{C}) = \mathcal{A}(\mathcal{O},\mathcal{C})\ :\ \mathcal{W} = \mathcal{B} \bigsqcup \mathcal{H}\,, \tag{3.2}$$

where

$$\mathcal{B} = \Phi^{-1}(\Lambda_\mathcal{B}),\ \mathcal{H} = \Phi^{-1}(\Lambda_\mathcal{H})\,. \tag{3.3}$$

## 3.2 "Voting by Elementary Features" Algorithms

To illustrate our classification approach, we introduce in this section some simple algorithms of the class "Voting by elementary features" [79, 85].

We symbolize by $a_\ell$ ($b_\ell$, $c_\ell$, $d_\ell$) — the number of objects $w \in \mathcal{B}_0$ (or $w \in \mathcal{H}_0$, or $w \in \mathcal{B}_0$, or $w \in \mathcal{H}_0$) such as $i$-th component of the vector $\omega = \Phi(w)$ is equal to 0 (or to 0, or to 1) :

$$r_\ell = \{\omega : \omega(\ell) = 0\}, \quad \overline{r_\ell} = \{\omega : \omega(\ell) = 1\} \tag{3.4}$$

$$\mathcal{R}_\ell = \Phi^{-1}(r_\ell) \bigcap \mathcal{W}_0, \quad \overline{\mathcal{R}_\ell} = \Phi^{-1}(\overline{r_\ell}) \bigcap \mathcal{W}_0 \,. \tag{3.5}$$

It is obvious that $\Omega_n = r_\ell \bigsqcup \overline{r_\ell}$, and $\mathcal{W}_0 = \mathcal{R}_\ell \bigsqcup \overline{\mathcal{R}_\ell}$.
For every $\ell = 1, \ldots, n$, the numbers $a_\ell$, $b_\ell$, $c_\ell$, $d_\ell$, that characterize the learning material and coding, are defined by two decompositions: $\mathcal{W}_0 = \mathcal{B}_0 \bigsqcup \mathcal{H}_0$ and $\mathcal{W}_0 = \mathcal{R}_\ell \bigsqcup \overline{\mathcal{R}_\ell}$. These numbers are: $a_\ell = \mid \mathcal{B}_0 \bigcap \mathcal{R}_\ell \mid$, $b_\ell = \mid \mathcal{H}_0 \bigcap \mathcal{R}_\ell \mid$, $c_\ell = \mid \mathcal{B}_0 \bigcap \overline{\mathcal{R}_\ell} \mid$, $d_\ell = \mid \mathcal{H}_0 \bigcap \overline{\mathcal{R}_\ell} \mid$.
The numbers $a_\ell/(a_\ell + c_\ell)$ and $b_\ell/(b_\ell + d_\ell)$ characterize the frequency of zeros in the $\ell$-th component of the learning objects from $\mathcal{B}_0$ and $\mathcal{H}_0$ correspondingly.
We introduce the vector $\mathbf{z} = (z(1), \ldots, z(n)) \in \Omega_n$. $z(1) = 0$, if $a_\ell/(a_\ell + c_\ell) \geq b_\ell/(b_\ell + d_\ell)$ and $z(1) = 1$ in the contrary case; $\ell = 1, \ldots, n$. The vector $\mathbf{z}$ is referred to as a **kernel** of the class $\mathcal{B}$.
This name signifies that the coordinates $z(\ell)$ have the values which are, in a certain sense, most characteristic for the learning class $\mathcal{B}_0$. After a simple transform we obtain:

$$\frac{a_\ell}{a_\ell + c_\ell} - \frac{b_\ell}{b_\ell + d_\ell} = \frac{1}{\mid \mathcal{B}_0 \mid \cdot \mid \mathcal{H}_0 \mid} \det \begin{pmatrix} a_\ell & b_\ell \\ c_\ell & d_\ell \end{pmatrix} \,. \tag{3.6}$$

Therefore,

$$z(\ell) = 0, \text{if} \det \begin{pmatrix} a_\ell & b_\ell \\ c_\ell & d_\ell \end{pmatrix} \geq 0 \tag{3.7}$$

and $z(\ell) = 1$ in the contrary case.
We introduce the following metric in the space $\Omega_n$:

$$\rho(\omega', \omega'') = \sum_{i=1}^{n} \mid \omega'(i) - \omega''(i) \mid \,. \tag{3.8}$$

Let us consider in the metric space $(\Omega_n, \rho)$ a ball $B[z,R] = \{\omega \in \Omega_n : \rho(\omega, z) \leq R\}$ with radius $R$ and center at the point $z$. Let $\Lambda_B = B[z, R]$. The algorithm VEF-0 represents $\Omega_n$ in the appearance $\Omega_n = \Lambda_B \bigsqcup \Lambda_H$, where $\Lambda_H = \Omega_n \setminus B[z,R]$ [90]. Here $R$ serves as a free parameter of the algorithm. In other words, the algorithm VEF-0 carries out the classification $\mathcal{A}(0,R)$ : $\mathcal{W} = \mathcal{B} \bigsqcup \mathcal{H}$, including to the class $\mathcal{B}$ the vectors which are sufficiently close to the kernel of the class $\mathcal{B}$. These vectors "look similar" to the vector $\mathbf{z}$ which "characterizes" the learning material. However, the degree of "similarity" for different coordinates remains equal.
In a natural way, the algorithm VEF-0 may be extended to decomposition into several classes. Let us illustrate this by an example of a decomposition into four classes. This case corresponds to the problem of recognizing zones of deep and shallow earthquakes. This algorithm, which we symbolize as VEF-04 [84], works in the following way.
Let us have the subsets $\mathcal{K}_0, \Gamma_0, \mathcal{H}_0, \subset \mathcal{W}$ which define the learning material. It is given that $\mathcal{H}_0 \bigcap \mathcal{K}_0 = \Gamma_0 \bigcap \mathcal{H}_0 = \emptyset$, while the sets $\mathcal{K}_0$ and $\Gamma_0$ may overlap. In this case, $\mathcal{H}_0 \bigcap \Gamma_0$ is the fourth learning class. Therefore, we have the following given decomposition $\mathcal{W} = (\Gamma_0 \bigcup \mathcal{K}_0) \bigsqcup \mathcal{H}_0$.
Let us introduce the following symbols:

| Number of objects from $\mathcal{W}$ | | such as $\omega(\ell) =$ |
|---|---|---|
| $a_\ell^K$ | $\mathcal{K}_0$ | 0 |
| $a_\ell^\Gamma$ | $\Gamma_0$ | 0 |
| $b_\ell^K$ | $\mathcal{W} \setminus \mathcal{K}_0$ | 0 |
| $b_\ell^\Gamma$ | $\mathcal{W} \setminus \Gamma_0$ | 0 |
| $c_\ell^K$ | $\mathcal{K}_0$ | 1 |
| $c_\ell^\Gamma$ | $\Gamma_0$ | 1 |
| $d_\ell^K$ | $\mathcal{W} \setminus \mathcal{K}_0$ | 1 |
| $d_\ell^\Gamma$ | $\mathcal{W} \setminus \Gamma_0$ | 1 |

In the first stage the algorithm VEF-04 constructs the sets $\mathcal{K}$ and $\Gamma$ which consist of the objects similar to those from $\mathcal{K}_0$ and $\Gamma_0$. Let $\mathcal{A}$ be equal to $\mathcal{K}$ or $\Gamma$.
Analogous to the algorithm VEF-0 we define the vector $\mathbf{Z}^\mathcal{A} = (z^\mathcal{A}(1), \ldots, z^\mathcal{A}(n)) \in \Omega_n$, to be a kernel of class $\mathcal{A}$ such that for $\ell = 1, \ldots, n$:

$$z^\mathcal{A}(\ell) = \begin{cases} 0 & \text{if } a_\ell^\mathcal{A}/(a_\ell^\mathcal{A} + c_\ell^\mathcal{A}) \geq b_\ell^\mathcal{A}/(b_\ell^\mathcal{A} + d_\ell^\mathcal{A}) \\ 1, & \text{in the contrarary case.} \end{cases} \tag{3.9}$$

We introduce the following determinants:

$$D_\ell^{\mathcal{K}} = \det \begin{pmatrix} a_\ell^{\mathcal{K}} & b_\ell^{\mathcal{K}} \\ c_\ell^{\mathcal{K}} & d_\ell^{\mathcal{K}} \end{pmatrix} \text{ and } D_\ell^{\Gamma} = \det \begin{pmatrix} a_\ell^{\Gamma} & b_\ell^{\Gamma} \\ c_\ell^{\Gamma} & d_\ell^{\Gamma} \end{pmatrix}, \ell = 1, \ldots, n. \tag{3.10}$$

It is obvious that we can define the kernel of class $\mathcal{A}$ as follows:

$$z^{\mathcal{A}} \in \Omega_n \text{is called kernel of class } \mathcal{A}, \text{ if } z^{\mathcal{A}}(\ell) = \begin{cases} 0, & \text{if } D_\ell^{\mathcal{A}} \geq 0 \\ 1, & \text{if } D_\ell^{\mathcal{A}} < 0 \end{cases}. \tag{3.11}$$

Let us construct the kernels $z^{\mathcal{K}}$, $z^{\Gamma}$ of classes $\mathcal{K}$ and $\Gamma$ and define in the space $\Omega_n$ the metric (3.8).
Introducing real numbers $R_{\mathcal{K}}$ and $R_{\Gamma}$ as free parameters of the algorithm, we consider the following two balls in the metric space $(\Omega_n, \rho)$,

$$\begin{array}{rcl} B[z^{\mathcal{K}}, R_{\mathcal{K}}] & = & \{\omega : \rho(\omega, z^{\mathcal{K}}) \leq R_{\mathcal{K}}\} \\ B[z^{\Gamma}, R_{\Gamma}] & = & \{\omega : \rho(\omega, z^{\Gamma}) \leq R_{\Gamma}\} \end{array}. \tag{3.12}$$

Let us define

$$\begin{array}{ll} \mathbf{K} = B[z^{\mathcal{K}}, R_{\mathcal{K}}] \setminus B[z^{\Gamma}, R_{\Gamma}] ; & \mathbf{\Gamma} = B[z^{\Gamma}, R_{\Gamma}] \setminus B[z^{\mathcal{K}}, R_{\mathcal{K}}] \\ \mathbf{K\Gamma} = B[z^{\mathcal{K}}, R_{\mathcal{K}}] \bigcap B[z^{\Gamma}, R_{\Gamma}] ; & \mathbf{H} = \Omega_n\{B[z^{\mathcal{K}}, R_{\mathcal{K}}] \bigcup B[z^{\Gamma}, R_{\Gamma}]\} \end{array} \tag{3.13}$$

Therefore, we obtain the decomposition of $\Omega_n$ into four non-overlapping classes

$$\Omega_n = \mathbf{K} \bigsqcup \mathbf{\Gamma} \bigsqcup \mathbf{K\Gamma} \bigsqcup \mathbf{H}$$

Applying the transform $\Phi^{-1}$, we obtain the following decomposition of the initial set of objects,

$$\mathbf{W} = \Phi^{-1}(\mathbf{K}) \bigsqcup \Phi^{-1}(\mathbf{\Gamma}) \bigsqcup \Phi^{-1}(\mathbf{K\Gamma}) \bigsqcup \Phi^{-1}(\mathbf{H}). \tag{3.14}$$

The last condition may be interpreted as a prediction, obtained by the algorithm VEF-04 on the basis of the learning material $(\mathcal{K}_0, \Gamma_0, \mathcal{H}_0)$.

The algorithm VEF-0 is a particular case of the following more general algorithm which we call VEF-1 [89, 90]. In VEF-1 "similarities" for different coordinates have, generally speaking, different "influence" to the classification. Formally speaking, this means the following.

Let us fix a vector $\sigma = (\sigma_1, \ldots, \sigma_n)$ with real components $\sigma_i \geq 0$ and introduce in $\Omega_n$ the metric, $\rho^\sigma$ defined as

$$\rho^\sigma(\omega', \omega'') = \sum_{i=1}^{n} \sigma_i \mid \omega'(i) - \omega''(i) \mid . \tag{3.15}$$

The same way as in the case of the algorithm VEF-0 we introduce the class $\mathcal{B} \subseteq \mathcal{W}$ to be the co-image of the ball of radius $R$ in metric $\rho^\sigma$ and with center in the kernel $z$. The algorithm VEF-1 gives the classification of the set of objects $\mathcal{W}$ in the appearance $\mathcal{W} = \mathcal{B} \bigsqcup \mathcal{H}$, where $\mathcal{H} = \mathcal{W} \setminus \mathcal{B}$. Thus, $R$ serves as a free parameter of the algorithm. It is obvious that VEF-0 is a particular case of VEF-1, if $\sigma = (1, 1, \ldots, 1)$.
Introducing the algorithm VEF-1 we assume that the vector $\sigma$ is fixed and does not depend on an initial decomposition $\mathcal{P}: \mathcal{W}_0 = \mathcal{B}_0 \bigsqcup \mathcal{H}_0$. It is natural, also, to consider the case when $\sigma = \sigma(\mathcal{P})$. In particular, we shall use below the following dependence of $\sigma$ on $\mathcal{P}$: components $\sigma_\ell = \sigma_\ell(\mathcal{P})$ are proportional to the modulus of the differences in frequencies:

$$\frac{a_\ell}{a_\ell + c_\ell} - \frac{b_\ell}{b_\ell + d_\ell}, \quad \sigma'_\ell = \det \begin{pmatrix} a_\ell & b_\ell \\ c_\ell & d_\ell \end{pmatrix}, \sigma_\ell = \mid \sigma_\ell \mid, \ell = 1, \ldots, n. \tag{3.16}$$

This version of the algorithm we call "VEF with flexible weights" and symbolize as VEF-2 [84].
One should be careful with the fact that if $a_\ell d_\ell - b_\ell c_\ell = 0$ we obtain that $\sigma_\ell = 0$ and $(\Omega_n, \rho)$ is, strictly speaking, not a metric space. However, this does not lead to any problem in what follows.
At the same time, when we apply the algorithms VEF-0, VEF-1 and VEF-2 to concrete applied problems, it is wise to change the numbers $a_i$, $b_i$, $c_i$, $d_i$ to the numbers $(a_i + 1)$, $(b_i + 1)$, $(c_i + 1)$, $(d_i + 1)$, to avoid an appearance of zero in the denominators. In other words, we introduce the following functions,

$$\mathcal{P}_i(\mathcal{B}_0) = \frac{1}{\mid \mathcal{B}_0 \mid + 2} \sum_{w \in \mathcal{B}_0} (\Phi(w))(i) + 1; \quad \mathcal{P}_i(\mathcal{H}_0) = \frac{1}{\mid \mathcal{H}_0 \mid + 2} \sum_{w \in \mathcal{H}_0} (\Phi(w))(i) + 1 . \tag{3.17}$$

The kernel of the class $\mathcal{B}$ for algorithms VEF-0, VEF-1 and VEF-2 in this case is defined by the condition

$$z(i) = \begin{cases} 1, & \mathcal{P}_i(\mathcal{B}_0) \geq \mathcal{P}_i(\mathcal{H}_0) \\ 0, & \mathcal{P}_i(\mathcal{B}_0) < \mathcal{P}_i(\mathcal{H}_0) \end{cases} . \tag{3.18}$$

## 3.3 Definition of Classes of Algorithms "Voting by a Set of Features" (VSF) and "Voting by Elementary Features" (VEF). Normal Weights

The algorithms described in section 3.2 are generalized by the following class of algorithms which we call VSF ("Voting by a Set of Features") [79, 84]. We now introduce this class.

To make the formulas more simple, we assume below that $\mathcal{W} = \mathcal{W}_0$. This assumption does not limit the constructions because a classification of an object may be different in different moments of time. Furthermore, consideration of such dynamical problems is the main goal of this part of the book.

A feature $\Upsilon$ is an arbitrary subset of $\Omega_n$. Let us consider a set of features $\mathcal{U} = \{\Upsilon_1, \ldots, \Upsilon_n\}$. We say that an object $w$ has a feature $\Upsilon$, if $\omega = \Phi(w) \in \Upsilon$. Each feature $\Upsilon$ defines the set $\Phi^{-1}(\Upsilon) \subseteq \mathcal{W}$ of objects which have the feature $\Upsilon$. We introduce the real vector of the feature weights $\sigma = (\sigma_1, \ldots, \sigma_n) = (\sigma(r_1, \mathcal{P}), \ldots, \sigma(r_n, \mathcal{P}))$ which, generally speaking, depends of an initial decomposition $\mathcal{P}$.

An algorithm of the class VSF determines whether an object belongs to the class $\mathcal{B}$ or $\mathcal{H}$ by voting on the set of features $\mathcal{U}$.

The voting $\hat{\mathcal{F}}(w)$ is defined as the sum of the weights of the features, which has an object $w$. The result of voting depends only of $\omega = \Phi(w)$:

$$\hat{\mathcal{F}}(w) = \mathcal{F}(\omega) = \sum_{\Upsilon \in \mathcal{U}\,:\,\Phi(w) \in \Upsilon} \sigma(\Upsilon, \mathcal{P}) \tag{3.19}$$

At the next step, a threshold of voting $\Delta$ is introduced. We include an object $w \in \mathcal{W}$ into the class $\mathcal{B}$, if $\hat{\mathcal{F}}(w) \geq \Delta$ and into the the class $\mathcal{H}$, if $\hat{\mathcal{F}}(w) < \Delta$. Therefore, to define a VSF algorithm, we need to fix a set of features $\mathcal{U}$ and a vector of weights $\sigma = \sigma(\mathcal{P})$. We shall symbolize this algorithm as VSF$(\mathcal{U}, \sigma(\mathcal{P}))$. Thus, $\Delta$ remains its only free parameter.

The applied problems under consideration below, lead to the necessity of considering specific types of functions $\sigma$ $(\mathcal{P})$. The function $\sigma(\Upsilon, \mathcal{P})$ is called "large", if "many" objects from $\mathcal{B}_0$ and "few" objects from $\mathcal{H}_0$ have the feature $\Upsilon$. To obtain such a type of function, we consider two decompositions of the set of objects $\mathcal{P}$:

$$\mathcal{W} = \mathcal{B}_0 \bigsqcup \mathcal{H}_0 \text{ and } \mathcal{W} = \Phi^{-1}(\Upsilon) \bigsqcup \overline{\Phi^{-1}(\Upsilon)}. \tag{3.20}$$

Analogous to section 3.2, we introduce the following:

$$\begin{array}{ll} a_\Upsilon = \mid \mathcal{B}_0 \bigcap \Phi^{-1}(\Upsilon) \mid, & b_\Upsilon = \mid \mathcal{H}_0 \bigcap \Phi^{-1}(\Upsilon) \mid \\ c_\Upsilon = \mid \mathcal{B}_0 \bigcap \overline{\Phi^{-1}(\Upsilon)} \mid, & d_\Upsilon = \mid \mathcal{H}_0 \bigcap \overline{\Phi^{-1}(\Upsilon)} \mid \end{array} \tag{3.21}$$

The weight defined by the formula:

$$\sigma(\Upsilon,\ \mathcal{P}) = \det \begin{pmatrix} a_\Upsilon & b_\Upsilon \\ c_\Upsilon & d_\Upsilon \end{pmatrix} = a_\Upsilon d_\Upsilon - c_\Upsilon b_\Upsilon\,, \tag{3.22}$$

is called the normal weight and is denoted by $\sigma^{\mathcal{H}}(\Upsilon,\ \mathcal{P})$. The vector $\sigma = (\sigma^{\mathcal{H}}(\Upsilon_1,\ \mathcal{P}),\ \ldots,\ \sigma^{\mathcal{H}}(\Upsilon_N,\ \mathcal{P}))$ is called the vector of normal weights of the set of features $\mathcal{U}$. We symbolize it by either $\sigma^{\mathcal{H}}$ or $\sigma^{\mathcal{H}}(\mathcal{P})$.

The case of normal weights will be studied below in detail. Here, we just mention that if $\Upsilon$ and $\overline{\Upsilon}$ are two complementary features (e.g. $\Omega_n = \Upsilon \bigsqcup \overline{\Upsilon}$), then $\sigma(\Upsilon) + \sigma(\overline{\Upsilon}) = 0$. Indeed, any object $w \in \mathcal{W}$ has one and only one of two features: $\Upsilon$ or $\overline{\Upsilon}$. Thus, $\overline{\Phi^{-1}(\Upsilon)} = \Phi^{-1}(\overline{\Upsilon})$, and the substitution $\sigma(\Upsilon)$ to $\sigma(\overline{\Upsilon})$ just exchanges the rows in the determinant.

Let us show that the algorithms VEF-0, VEF-1 and VEF-2 are included in the class VSF for some values of the weights, $N = n$, and $\mathcal{U} = \mathcal{U}_\Gamma = \{\Upsilon_1,\ \ldots,\ \Upsilon_N\}$, where $\Upsilon_i = \{w:\ w(i) = 0\}$, $i = 1,\ \ldots,\ n$. In fact, the function of voting 3.19 in this case has the appearance

$$\mathcal{F}(\omega) = \sum_{i=1}^{n} \sigma_i (1 - \omega(i))\,, \tag{3.23}$$

where we introduce the following symbols:

$$\sigma_+ = (\mid \sigma_1 \mid,\ \ldots,\ \mid \sigma_n \mid);\ \mathbf{sgn}(\sigma) = (sgn(\sigma_1),\ \ldots,\ sgn(\sigma_n))\,,$$

with,

$$sgn(x) = 1,\ \text{if } x \geq 0;\ sgn(x) = 0,$$

if $x < 0$; $\overline{0} = (0,\ \ldots,\ 0)$, $e = (1,\ \ldots,\ 1) \in \Omega_n$, $\overline{w} = e - w$.
It is clear that the function $\mathcal{F}$ has its maximum value at the point $z = e - sgn(\sigma) \in \Omega_n$, and minimum value at the point $\overline{z} = sgn(\sigma) \in \Omega_n$. We

have, also, $\mathcal{F}(z) - \mathcal{F}(\overline{z}) = \sum_{i=1}^{n} \mid \sigma_i \mid$. In particular, when all the weights are non-negative, we have $z = 0, \mathcal{F}(z) = \sum_{i=1}^{n} \sigma_i$; $\overline{z} = e$; $\mathcal{F}(\overline{z}) = 0$. Therefore, the threshold of voting $\Delta$ should be chosen in the interval $\mathcal{F}(\overline{z}) \leq \Delta \leq \mathcal{F}(z)$. The domain $\mathcal{F}(w) = \sum_{i=1}^{n} \sigma_i(1 - \omega(i)) \geq \Delta$ in $\Omega_n$ we call, as usual, the half-space. The following statement shows that this half-space is a ball in a metric $\rho^{\sigma+}$:

**Statement** *Let the function $\mathcal{F}(\omega)$ be defined by the formula 3.23. Then the set $\{w \in \Omega_n : \mathcal{F}(\omega) \geq \Delta\}$ in a metric space $(\Omega_n, \rho^{\sigma+})$ is a ball of radius $\mathcal{F}(z) - \Delta$ with the center in the point $z = e - sgn(\sigma)$.*

## Proof

From the equation 3.23 we obtain the following formula:

$$\mathcal{F}(z) - \mathcal{F}(\omega) = \sum_{i=1}^{n} \sigma_i(\omega(i) - z(i)) = \sum_{i=1}^{n} \mid \sigma_i \mid \mid z(i) - \omega(i) \mid = \rho^{\sigma+}(\omega, z). \tag{3.24}$$

Therefore, the condition $\mathcal{F}(\omega) \geq \Delta$ is equivalent to the condition $\rho^{\sigma+}(w, z) \leq \mathcal{F}(z) - \Delta$, which proves the statement.
In the same way, we may prove that the condition $\mathcal{F}(\omega) < \Delta$ defines in the space $(\Omega_n, \rho^{\sigma+})$ the ball of radius $\Delta - \mathcal{F}(\overline{z}) - \epsilon$ with center at the point $\overline{z}$. Here, $\epsilon$ is any sufficiently small real positive number. Thus , the condition $\mathcal{F}(\omega) \geq \Delta$ defines the decomposition of $\Omega_n$ into two non-overlapping balls with centers at the points $z$ and $\overline{z}$ and with the sum of radii equal to $\sum_{i=1}^{n} \mid \sigma_i \mid - \epsilon$.
As it was shown in section 3.2, the algorithms VEF-0, VEF-1 and VEF-2 give the classification $\Omega_n = \Lambda_{\mathcal{B}} \bigsqcup \Lambda_{\mathcal{H}}$, where $\Lambda_{\mathcal{B}}$ is a ball in $\Omega_n$ with the center in the kernel $z = e - sgn(\sigma^{\mathcal{H}})$ and with metric $\rho^{\mathcal{C}}$, $\mathcal{C} = (C_1, \ldots, C_n)$, $C_i > 0$. We have that $\mathcal{C} = \sigma_+^{\mathcal{H}}$ for VEF-2, $\mathcal{C} = const$ for VEF-1, $\mathcal{C} = (1, \ldots, 1)$ for VEF-0. Therefore, if we take the vector of normal weights $\sigma = \sigma^{\mathcal{H}}$, we obtain that VSF( $\mathcal{U}_\Gamma$, $\sigma^{\mathcal{H}}(\mathcal{P})$) = VEF-2. If we consider $\sigma_i = \sigma_i^{\mathcal{H}} \mid \sigma_i^{\mathcal{H}} \mid^{-1}$, then VSF( $\mathcal{U}_\Gamma$, $\sigma(\mathcal{P})$) = VEF-0. Finaly, if we take $\sigma_i = C_i\sigma_i^{\mathcal{H}} \mid \sigma_i^{\mathcal{H}} \mid^{-1}$, then VSF( $\mathcal{U}_\Gamma$, $\sigma(\mathcal{P})$) = VEF-1. In all three cases the point $z = e - sgn(\sigma)$ is equal to the kernel of the class $\mathcal{B}$.
Generalizing this result, we define an algorithm of the class VEF to be all algorithms VSF( $\mathcal{U}_\Gamma$, $\sigma(\mathcal{P})$), that have the set of features $\mathcal{U}$ equal to $\mathcal{U}_\Gamma$, and an arbitrary vector of weights $\sigma(\mathcal{P})$. Due to the statement, classifications obtained using algorithms of the class VEF have an appearance $\Omega_n =$

$\Lambda_{\mathcal{B}} \bigsqcup \Lambda_{\mathcal{H}}$, where $\Lambda_{\mathcal{B}}$, $\Lambda_{\mathcal{H}}$ are the balls in $(\Omega_n,\ \rho^{\sigma+})$ with the centers at the points $z = e - sgn(\sigma)$, $\overline{z} = sgn(\sigma)$ and radii $\mathcal{F}(z) - \Delta$, $\Delta - \mathcal{F}(\overline{z}) - \epsilon$.

## 3.4 Class of Algorithms "CORA-i"

The algorithm CORA was introduced by M. Bongard and M. Vainzveig in the early 1970's [22, 177]. It was used by the present authors as well as by many other researchers in a wide circle of applied problems. One of the examples pertains to mineral deposit exploration. In the domain of CORA-3 are applications to the problems of the classification of oil bearing and water bearing layers [22] and exploration of uranium deposits [27]. An extended project of pattern recognition application to medical diagnostics based on this algorithm was developped by I. Gelfand and his collaborators in 70's and early 80's [70].
In the late 1970's, L. Knopoff, F. Press, I. Gelfand and V. Keilis-Borok started the project of applying CORA-3 to the problem of recognition of strong earthquake-prone areas [66, 67, 68, 69]. In the 1980's, the algorithm was used by the authors of this book and, A. Cisternas, C. Weber, H. Philip, V. Kossobokov, A. Soloviev, E. Ranzman *et al* [31, 80, ?, 92, 142, 178, 179] as one of the approaches in a French-Russian project of determining earthquake-prone areas in regions of moderate seismicity.

We describe the algorithm CORA-3 in its classical appearance. We symbolize this realization of CORA 3 as $\aleph$.
A feature is defined by the following matrix,

$$\tau = \begin{pmatrix} i_1 & i_2 & i_3 \\ \alpha_1 & \alpha_2 & \alpha_3 \end{pmatrix} \tag{3.25}$$

where $i_r = 1, \ldots, n$; $r = 1, 2, 3$; $\alpha_r = 0$ or $\alpha_r = 1$.
In other words, by "feature" we mean a value of a coordinate, their pairs and thirds, supplied with the corresponding coordinate numbers.
Vector $\omega \in \Omega_n$ has a feature $\tau$, if $\omega(i_r) = \alpha_r, r \leq 3$. Let $\mathcal{Y} = \Phi(\mathcal{B}_0)$ (or $\mathcal{Y} = \Phi(\mathcal{H}_0)$) be the set of vectors which corresponds to the learning material. By $K_{\mathcal{Y}}(\tau)$ we mean the number of vectors $\omega \in \mathcal{Y}$ which have a given feature $\tau$.
Then, we introduce four free natural parameters: two thresholds of selection, $K_{\mathcal{B}}$, $K_{\mathcal{H}}$, and two thresholds of rejection, $\tilde{K_{\mathcal{B}}}$, $\tilde{K_{\mathcal{H}}}$.

A feature $\tau$ is called characteristic for the class $\mathcal{B}$, if it appears sufficiently "often" in the learning material $\Phi(\mathcal{B}_0)$, $K_{\Phi(\mathcal{B}_0)}(\tau) \geq K_{\mathcal{B}}$, $K_{\Phi(\mathcal{H}_0)}(\tau) \leq \tilde{K_{\mathcal{B}}}$. Correspondingly, a feature $\tau$ is called characteristic for the class $\mathcal{H}$, if $K_{\Phi(\mathcal{H}_0)}(\tau) \geq K_{\mathcal{H}}$, $K_{\Phi(\mathcal{B}_0)}(\tau) \leq \tilde{K_{\mathcal{H}}}$.
In the set of characteristic features of the class $\mathcal{B}$ (or $\mathcal{H}$) we introduce the following relation of equivalence: $\tau \sim \nu$, if these features appear as the same vectors from $\Phi(\mathcal{B}_0)$ (or from $\phi(\mathcal{H}_0)$). We also introduce the relation of a particular order assuming that a feature $\tau$ is subordinated to a feature $\nu$, if every vector $\omega \in \Phi(\mathcal{B}_0)$ (correspondingly $\omega \in \Phi(\mathcal{H}_0)$) which has the feature $\tau$, also has the feature $\nu$.
On the basis of the sets of characteristic features of the classes $\mathcal{B}$ and $\mathcal{H}$, we construct their subsets, which include only one feature from each equivalence class. Then, we exclude from these subsets all the features which are subordinated to others. We denote the subsets thus obtained as $\mathcal{R}_{K_{\mathcal{B}},\, \tilde{K_{\mathcal{B}}}}(\mathcal{B}_0)$ and $\mathcal{R}_{K_{\mathcal{H}},\, \tilde{K_{\mathcal{H}}}}(\mathcal{H}_0)$.
By forming the sets $\mathcal{R}_{K_{\mathcal{B}},\tilde{K_{\mathcal{B}}}}(\mathcal{B}_0)$ and $\mathcal{R}_{K_{\mathcal{H}},\, \tilde{K_{\mathcal{H}}}}(\mathcal{H}_0)$ we are completing the stage of learning. On the next stage, we examine the objects $\omega \in \mathcal{W}$ by voting.
In the main version of the algorithm the following simple logical voting model is used. For every vector $\omega \in \Omega_n$ we calculate $\Delta(\omega) = N_{\mathcal{B}}(\omega) - N_{\mathcal{H}}(\omega)$, where $N_{\mathcal{B}}(\omega)$ (or $N_{\mathcal{H}}(\omega)$) is the number of features from $\mathcal{R}_{K_{\mathcal{B}}\, \tilde{K_{\mathcal{B}}}}(\mathcal{B}_0)$, (or from $\mathcal{R}_{K_{\mathcal{H}},\, \tilde{K_{\mathcal{H}}}}(\mathcal{H}_0)$) which have the vector $\omega$. Then, we introduce one more natural parameter $\Delta$ which is called the voting threshold. The classifications produced by the algorithm $\aleph(\mathcal{O};\, K_{\mathcal{B}},\, \tilde{k_{\mathcal{B}}},\, K_{\mathcal{H}},\, \tilde{K_{\mathcal{H}}},\, \mu,\, \Delta)$ (here $\mathcal{O} = (\mathcal{B}_{\prime},\, \mathcal{H}_{\mathcal{O}})$, and $\mu$ is the parameter which corresponds to the freedom of the choice of a parameter from the aquivalence class), have the appearence

$$\begin{aligned} \mathcal{B} = &\ \{\omega \in \mathcal{W}:\ \Delta(\Phi(\omega)) \ \geq \Delta\} \\ \mathcal{H} = &\ \{\omega \in \mathcal{W}:\ \Delta(\Phi(\omega)) \ < \Delta\} \end{aligned} \tag{3.26}$$

The algorithm $\aleph$ is also included in the class VSF. To show this, we need the following:

**Definition**. We define a face to be a subset $\Gamma \subset \Omega_n$ which consists of all the points $\omega = (\omega(1),\ \ldots,\ \omega(n))$, for which the coordinates with numbers $i_1,\ \ldots,\ i_K$ are arbitrary (e.g. equal to 0 or 1), while the other $(n-K)$ coordinates are fixed. We say that a face has the dimension $K$ and is defined by the edges of length $\sigma_{i_1},\ \ldots,\ \sigma_{i_K}$.

An algorithm $\aleph$ is included to the class VSF if we take as a set of features $\mathcal{U}_{\aleph_3}$ all the faces of $\Omega_n$ which have the dimensions $(n-1)$, $(n-2)$, $(n-3)$. In fact, let us fix a feature $\Upsilon \in \mathcal{U}_{\aleph_3}$ and calculate the coefficients $a_\Upsilon$, $b_\Upsilon$, $c_\Upsilon$, $d_\Upsilon$ using the formula (3.2.1). Then it is not complicated to check that if we introduce the weights by the formula

$$\sigma(\Upsilon,\, \mathcal{P}) = sgn(a_\Upsilon - K_\mathcal{B})sgn(\tilde{K_\mathcal{B}} - b_\Upsilon) - sgn(\tilde{K_\mathcal{H}} - a_\Upsilon)sgn(b_\Upsilon - K_\mathcal{H}) \quad (3.27)$$

we obtain that $\mathrm{VSF}(\mathcal{U}_{\aleph_3},\, \sigma(\mathcal{P})) = {\aleph_\ni}'$, where $\aleph'$ differs from $\aleph$ only by the fact that we take into consideration all the equivalent and subordinated features. In other words, in the algorithm $\aleph'$ the weights have three values:

- $+1$, for the characteristic features of the class $\mathcal{B}$,
- $-1$, for the characteristic features of the class $\mathcal{H}$,
- $0$, for all other features.

Shifting ourselves from $\aleph'$ to $\aleph$ we give zero weights to all induced features. Generalizing the result obtained, we symbolize by CORA-i an algorithm $\mathrm{VSF}(\mathcal{U}_{\aleph_i},\, \sigma(\mathcal{P}))$, where $\mathcal{U}_{\aleph_i}$ are the faces of $\Omega_n$ having dimensions $(n-1)$, $(n-2)$, $\ldots$, $(n-i)$. Note that the set of features $\mathcal{U}_{\aleph_i}$ includes the set $\mathcal{U}_\Gamma$ (see section 3.3). Thus, the class of algorithms CORA-i contains all the algorithms of the class VEF (see table 1).

## 3.5 "Neighbours" as a VSF Algorithm

The widely used algorithm "Neighbours" is oriented toward problems where the learning set has the property $\mathcal{W}_0 \neq \mathcal{W}$. Different applications of this algorithm are described in [25, 59, 146, 160, 175, 183, 184, 185, 186].
The algorithm "Neighbours" functions in the following way.
We consider the mapping $\Phi : \mathcal{W} \to \Omega$ of the set of objects to the space of parameter values. At the same time, we introduce in $\Omega$ a metric $\rho$. We consider below the case when $\Omega = \Omega_n$ and $\rho = \rho^\sigma$ (see formula 6.15). Objects from $\omega \in \mathcal{B}_0$ are classified into the class $\mathcal{B}$ and objects from $\mathcal{H}_0$ into class $\mathcal{H}$ by definition. For the rest of the objects $\omega^\star \in \mathcal{W} - \mathcal{W}_0$, the question about their classification is decided by voting. The objects $\omega \in \mathcal{W}_0$ take part in the voting. The objects $\omega \in \mathcal{B}_0$ vote for inclusion to the class $\mathcal{B}$. The objects $\omega \in \mathcal{H}_0$ vote against inclusion to the class $\mathcal{B}$.

Every object $\omega \in \mathcal{W}_0$ votes by its weight which monotonously decreases due to the decrease of the distance $\rho^\sigma(\Phi(\omega), \Phi(\omega^\star))$. More precisely $\omega^\star \in \mathcal{B}$ if, and only if,

$$\mathcal{F}_0(\omega^\star) = \sum_{\omega \in \mathcal{W}_0} \delta(\omega) f[\rho^\sigma(\Phi(\omega^\star), \Phi(\omega))] \geq \Delta\,, \tag{3.28}$$

where $\delta(\omega) = +1$, if $\omega \in \mathcal{B}_0$ and $\delta(\omega) = -1$, if $\omega \in \mathcal{H}_0$ and where $f$ is a monotonously decreasing function of non-negative argument, while $\Delta$ is a free parameter of the algorithm.

Now, we shall proof that "Neighbours" is also a VSF algorithm. Let us take as a set of features $\mathcal{U} = \mathcal{U}_c$ the set of all balls $\mathcal{B}(z, R) \in (\Omega_n, \rho^\sigma)$. We consider two balls $\mathcal{B}(z_1, R_1), \mathcal{B}(z_2, R_2)$ which are different if $z_1 \neq z_2$ and equal if $z_1 = z_2$ eventhough $R_1 \neq R_2$. The number of different balls is finite since $\Omega_n$ is a finite set. We look for a weight of a feature $\Upsilon = \mathcal{B}(z, R)$ in the form $\sigma\Upsilon = h(z)g(R)$ where $h(z) = \mid \Phi^{-1}(z) \bigcap \mathcal{B}_0 \mid - \mid \Phi^{-1}(z) \bigcap \mathcal{H}_0 \mid$ is the difference in the number of objects from $\mathcal{B}_0$ and $\mathcal{H}_0$ which $\Phi$ transfers to $z$, while $g$ is a function of positive argument. Let us show that we may define $g$ in a way such that algorithm VSF $(\mathcal{U}_c, \sigma)$ is equal to the algorithm "Neighbours".

In the space $(\Omega_n, \rho^\sigma)$ the distance have the finite number of values $\rho_0 = 0,\ \rho_1,\ \ldots,\ \rho_M = \sum \sigma_i$. Because of the symmetry of the space $\Omega_n$, for every point $z$ and $\omega$ we have $\rho(z, \omega) = \rho_{i_0}$. While voting for $\omega$ the point $z$ gives the contribution: $h(z)f(\rho_{i_0})$ in the algorithm "Neighbours" and contribution $(\mathcal{U}_c, \sigma) - h(z)\sum_{i \geq i_0} g(\rho_{i_0})$, in the algorithm VSF.

Thus, it is sufficient to define the function $g$ in such a way that

$$f(\rho_{i_0}) = \sum_{i \geq i_0} g(\rho_i), \quad \forall i = 0, \ldots, M\,. \tag{3.29}$$

We satisfy such an equation if we take

$$g(\rho_M) = f(\rho_M), \quad g(\rho_i) = f(r_i) - f(r_{i+1}), \quad i = 1, \ldots, M-1\,. \tag{3.30}$$

The obtained algorithm VSF$(\mathcal{U}_c, \sigma)$ and the algorithm "Neighbours" classify in the same way all the objects (except possibly the objects from $\mathcal{W}_0$) because the algorithm "Neighbours" does not classify these objects at all. We have the precise identity of the algorithms in the case of injective mapping $\Phi : \mathcal{W}_0 \to \Omega_n$ if $g(\rho_0)$ is taken to be a sufficiently large constant.

## 3.6 "Threshold" Classification Algorithms

Here we introduce a class of recognition algorithms which is more general than VSF. Consider a function $F : \Omega_n \to \mathcal{R}$, where $\mathcal{R}$ is the set of real numbers. We correspond to this function an algorithm $T(F)$, which defines the classification,

$$\Omega_n = \Lambda_B \bigsqcup \Lambda_H, \text{ where } \Lambda_B = \{\omega : F(\omega) \geq \Delta\}, \; \Lambda_H = \{\omega : F(\omega) < \Delta\}, \tag{3.31}$$

where $\Delta$ is a free parameter of the algorithm. As it follows from 3.3, the algorithm of the class VSF and in particular of the class VEF are "threshold" classification algorithms.
Threshold algorithms $T(F_1)$ and $T(F_2)$ are called equivalent if the functions $F_1$ and $F_2$ are connected by a monotonous transform; e.g., $F_1(\omega) = G(F_2(\omega))$ where $G$ is a strictly increasing function. It is clear that the classifications obtained by equivalent algorithms $T(F_1)$ and $T(F_2)$ are equal if $\Delta_1 = G(\Delta_2)$.
Any algorithm $T(F)$ with a linear function

$$F(\omega) = a + \sum_{i=1}^{n} c\omega(i) \tag{3.32}$$

is equivalent to some algorithm of the class VEF. In fact, according to the formula 3.23, an algorithm of the class VEF is a "threshold" algorithm $T(F_\Gamma)$ where

$$F_\Gamma(\omega) = \sum_{i=1}^{n} \sigma_i(i) - \sum_{i=1}^{n} \sigma_i \omega(i). \tag{3.33}$$

If $\sigma_i = -c_i$, $i = 1, \ldots, n$, then

$$F_\Gamma(\omega) = -\sum_{i=1}^{n} c_i + \sum_{i=1}^{n} c_i \omega(i). \tag{3.34}$$

Thus, $F$ and $F_\Gamma$ are connected by a strictly monotonous transform $F = G(F_\Gamma)$ where $G(x) = x + (a + \sum_{i=1}^{n} c_i)$.
Let the function $F : \Omega \to \mathcal{R}$ have an appearance

$$F(\omega) = \sum_{i=1}^{n} \mathcal{F}_i(\omega(i)), \tag{3.35}$$

where the $\mathcal{F}_i$ are arbitrary functions which realize a mapping $(0,1) \to \mathcal{R}$. We now show that the corresponding threshold algorithm $T(F)$ is also equivalent to some algorithm of the class VEF. Let us introduce $\sigma_i = \mathcal{F}_i(1) - \mathcal{F}_i(0)$. Then 3.35 can be represented in the appearance

$$F(\omega) = \sum_{i=1}^{n} \mathcal{F}_i(0) + \sum_{i=1}^{n} \sigma_i \omega(i). \tag{3.36}$$

Therefore, $F$ is a linear function over $\Omega_n$ and we may refer to the previous statement.

In the context of the introduced notion of equivalent algorithms we have the following correspondance between the classes of algorithms "Cora-i" and VEF: "Cora-1" with normal weights is equivalent to VEF-2.

To prove this, we note that, according to section 3.4

$$\mathcal{U}_{\mathcal{H}_1} = \{r_1, \ldots, r_n, \quad \overline{r_1}, \ldots, \overline{r_n}\} \tag{3.37}$$

where

$$r_i = \{\omega : \omega(i) = 0\}, \quad \overline{r_i} = \{\omega : \omega(i) = 1\}.$$

At the same time, as is shown in 3.3, $\sigma^H(r_i) = -\sigma^H(\overline{r_i})$, $i = 1, \ldots, n$. Applying formula 3.32 for the case of "Cora-1", we obtain that $F(\omega)$ is a sum of $n$ members: if $\omega(i) = 0$, then $\sigma_i^H$ is included in the sum, if $\omega(i) = 1$ then $-\sigma_i^H$ is included.

Therefore,

$$F(\omega) = F_{\mathcal{H}_1}(\omega) = \sum_{i=1}^{n} \sigma_i^H (1 - 2\omega(i)) = \sum_{i=1}^{n} \sigma_i^H - 2 \sum_{i=1}^{n} \sigma_i^H \omega(i). \tag{3.38}$$

At the same time, for VEF, formula 3.32 gives,

$$F(\omega) = F_{\Gamma_2}(\omega) = \sum_{i=1}^{n} \sigma_i^H - \sum_{i=1}^{n} \sigma_i^H \omega(i). \tag{3.39}$$

It is clear that $F_{\mathcal{H}_1}$ and $F_{\Gamma_2}$ are connected by the monotonous transformation,

$$F_{\mathcal{H}_1} = 2F_{\Gamma_2} - \sum_{i=1}^{n} \sigma_i^H. \tag{3.40}$$

## 3.7 "Bayes" as a VEF Algorithm

Bayes approach is widely used in many different problems of pattern recognition (see for example [89, 90, 158]). This approach is realized by the algorithm, which we shall call below as Bayes and symbolize by $\mathcal{B}$.
"Bayes" is a threshold algorithm. Let us prove it by introducing necessary formulas and giving them a probabilistic interpretation. We assume that the objects $w \in \mathcal{W}$ are choosen randomly with equal probability. We also assume that we have a map $\Phi: \mathcal{W} \to \Omega_n$ and a fixed vector $\omega_0 \in \Omega_n$.
We define $F(\omega_0)$ as a conditional probability

$$F(\omega_0) = \mathcal{P}\{\Phi(w) \in \Lambda_B \mid w = \omega_0\} . \tag{3.41}$$

Let us introduce the following events:

$$A_B: \Phi(w) \in \Lambda_B, \quad A_H: \Phi(w) \in \Lambda_H, \quad K: \Phi(w) = \omega_0$$

. Then using Bayes formula we represent $F(\omega_0)$ in the following way:

$$F(\omega_0) = \mathcal{P}\{A_B \mid K\} = \frac{\mathcal{P}\{K \mid A_B\}\mathcal{P}\{A_B\}}{\mathcal{P}\{K\}}$$

$$= \frac{\mathcal{P}\{K \mid A_B\}\mathcal{P}\{A_B\}}{\mathcal{P}\{K \mid A_B\}\mathcal{P}\{A_B\} + \mathcal{P}\{K \mid A_H\}\mathcal{P}\{A_H\}}$$

$$= \left[1 + \frac{\mathcal{P}\{A_H\}}{\mathcal{P}\{A_B\}}\left[\frac{\mathcal{P}\{K \mid A_B\}}{\mathcal{P}\{K \mid A_H\}}\right]^{-1}\right]^{-1} . \tag{3.42}$$

The probabilities in the last formula are estimated on the basis of an initial decomposition $P: W = B_0 \bigsqcup H_0$.
It is natural to assume that

$$\mathcal{P}\{A_B\} = \mid B_0 \mid / \mid W \mid, \quad \mathcal{P}\{A_H\} = \mid H_0 \mid / \mid W \mid . \tag{3.43}$$

At the same time, as it will be shown below, a classification defined by the algorithm $\mathcal{B}$ does not depend on $\mathcal{P}\{A_B\}$ and $\mathcal{P}\{A_H\}$.
Let us calculate the probabilities $\mathcal{P}\{K \mid A_B\}$ and $\mathcal{P}\{K \mid A_H\}$. To do so, we consider the events:

$$K_i^0: \omega(i) = 0 \text{ and } K_i^1: \omega(i) = 1 .$$

It is clear that

$$K = \bigcap_{i=1}^{n} K_i^{\omega_0(i)}.$$

The following conditional probabilities are defined from the corresponding frequencies:

$$\mathcal{P}\{K_i^0 \mid A_B\} = \frac{a_i}{a_i + c_i} = \alpha_i^0; \quad \mathcal{P}\{K_i^1 \mid A_B\} = \frac{c_i}{c_i + d_i} = \alpha_i^1$$

$$\mathcal{P}\{K_i^0 \mid A_H\} = \frac{b_i}{b_i + d_i} = \beta_i^0; \quad \mathcal{P}\{K_i^1 \mid A_H\} = \frac{d_i}{b_i + d_i} = \beta_i^1, \qquad (3.44)$$

where $a_i$, $b_i$, $c_i$, $d_i$ are defined in 3.2. Let us assume that the events $K_i^{\omega_0(i)} \bigcap A_B$ and $K_i^{\omega_0(i)} \bigcap A_H$ are independent for different $i = 1, \ldots, n$. Accepting this assumption we obtain,

$$\mathcal{P}\{K \mid A_B\} = \sqcap_{i=1}^{n} \alpha_i^{\omega_0(i)}; \quad \mathcal{P}\{K \mid A_H\} = \sqcap_{i=1}^{n} \beta_i^{\omega_0(i)}. \qquad (3.45)$$

Taking these formulas into account for $F(\omega_0)$ we obtain

$$F(\omega_0) = \left[1 + \lambda \left[\sqcap_{i=1}^{n} \frac{\alpha_i^{\omega_0(i)}}{\beta_i^{\omega_0(i)}}\right]^{-1}\right]^{-1}, \qquad (3.46)$$

where

$$\lambda = \frac{\mathcal{P}\{A_H\}}{\mathcal{P}\{A_B\}} > 0.$$

We symbolize

$$F'(\omega_0) = \sqcap_{i=1}^{n} \frac{\alpha_i^{\omega_0(i)}}{\beta_i^{\omega_0(i)}}. \qquad (3.47)$$

It is not complicated to see that the functions $F$ and $F'$ are connected by a strictly monotonous transform $F = G(F')$ where $G(x) = 1/(1 + \lambda/x)$. We introduce the function

$$F''(\omega_0) = \ln(F'(\omega_0)) = \sum_{i=1}^{n} \ln \frac{\alpha_i^{\omega_0(i)}}{\beta_i^{\omega_0(i)}}. \qquad (3.48)$$

The function $F''$ is connected with $F'$ (and thus with $F$) by the strictly monotonous transform $G(x) = \ln x$ and has the appearance 3.35. In other words,

$$F'' = \sum_{i=1}^{n} \mathcal{F}_i(\omega(i)), \text{ where } \mathcal{F}_i = \ln \frac{\alpha_i^{\omega_0(i)}}{\beta_i^{\omega_0(i)}}$$

depends only on $\omega(i)$. Therefore, according to 3.6, the algorithm $\mathcal{B}$ is equivalent to an algorithm VEF with the weights $\sigma_i = \mathcal{F}_i(1) - \mathcal{F}_i(0)$. Furthermore, because

$$\mathcal{F}_i(0) = \frac{\alpha_i^0}{\beta_i^0} = \ln \left[ \frac{a_i}{a_i + c_i} : \frac{b_i}{b_i + d_i} \right]$$

$$\mathcal{F}_i(1) = \frac{\alpha_i^1}{\beta_i^1} = \ln \left[ \frac{c_i}{d_i + c_i} : \frac{d_i}{b_i + d_i} \right] \tag{3.49}$$

we obtain:

$$\sigma_i = \mathcal{F}_i(1) - \mathcal{F}_i(0) = \ln \frac{a_i d_i}{b_i c_i}. \tag{3.50}$$

The weights $\sigma_i = \ln(a_i d_i) - \ln(b_i c_i)$ are not normal ones. However, they have some of their features. In particular, $sgn(\sigma_i) = sgn(\sigma_i^H)$.

**Remark**

If an initial decomposition $P: W = B_0 \bigsqcup H_0$ produces zero values of some of the numbers $a_i$, $b_i$, $c_i$, $d_i$ $(i = 1, \ldots, n)$, then the formula 3.50 for $\sigma_i$ has no sense. In this case we make the change

$$a_i \mapsto (a_i + 1), \; b_i \mapsto (b_i + 1), \; c_i \mapsto (c_i + 1), \; d_i \mapsto (d_i + 1).$$

In particular in this form the algorithm $\mathcal{B}$ was applied to the problem of recognition of **earthquake-prone areas** [89, 90].

# Chapter 4

# Dynamic and Limit Classification Problems

In this chapter we introduce a class of recognition problems which are characterized by the fact that the considered classifications of objects are time dependent. This fact gives an opportunity to formulate the condition of stability, which allows to restrict a set of classifications under consideration [79, 85, 87]. Thus, we obtain a new instrument to make the right choice of solution of a recognition problem among the set of classifications obtained by the family of algorithms described in Chapter 3. As we shall see in chapter 6, the limit case of dynamical recognition problem corresponds to the problem of recognition of strong earthquake prone-areas.

## 4.1 Formulation of the Dynamic Recognition Problem

Dynamic recognition problems are characterized by the variation in time of the process of the division of the set of objects into a finite number of classes. Let us give a formal definition.

Let $W$ be a finite set of objects and $t \in [T_0, T_1]$ an evolution parameter which has a sense of time $(T_1 < \infty)$. We assume that in every moment $t \in [T_0, T_1]$ there is a decomposition $P(t)$ of the set $W$ into $r$ non-overlapping classes.

$$p(t)\colon W = \bigsqcup_{i=1}^{r} W_i(t)\,, \tag{4.1}$$

(it is possible that some classes $W_i(t)$ are empty).
In other words, we have the mapping

$$\varphi : W \times [T_0, T_1] \to \{1, \ldots, r\}\,. \tag{4.2}$$

According to this mapping and the time evolution, the objects $w \in W$ are included into different classes at different moments in time. However, not all movements of an object from one class to another are possible. The allowed movements of the objects are defined by a graph $\Gamma$. The arrows of $\Gamma$ are marked by the integers $1, 2, \ldots, r$. Some of the arrows are connected by the oriented segments $[i,j]$, $i,j = 1, 2, \ldots, r$. A movement of an object from the class $i$ to the class $j$ is impossible, if $\Gamma$ does not contain the segment $[i,j]$. We call the graph $\Gamma$ as the graph of allowed movements.
Every moment in time, the objects $w \in W$ are characterized by the values of the set of parameters. In other words, we have the mapping

$$\Phi\colon W \times [T_0, T_1] \to \Omega\,, \tag{4.3}$$

where $\Omega$ is the space of parameters values.

Let us fix the momentin which we consider a dynamic pattern recognition problem. The known information, in this case, consists of the data on the behaviour of the system in the time interval $[T_0, t]$. Formally, this means that the functions $\varphi$ and $\Phi$ are known for some fixed subsets $R_\varphi(t)$, $R_\Phi(t) \subset W \times [T_0, t]$ correspondingly.

$$\text{The functions } \varphi \Big|_{R_\varphi(t)} \text{ and } \Phi \Big|_{R_\Phi(t)}$$

we call as learning material at the time moment $t$. We symbolize the learning material as

$$\mathcal{O}(t) = \left( \varphi \Big|_{R_\varphi(t)}, \Phi \Big|_{R_\Phi(t)} \right). \tag{4.4}$$

Let us fix a different moment of the time $t'$ such as $t < t' < T_1$. By a solution $\Pi(t,t')$ of a dynamic recognition problem we will mean the definition of classification $W = \bigsqcup_{i=1}^{r} W_i(t')$ by a pattern recognition algorithm (for example by a VSF algorithm) based on the learning material $\mathcal{O}(t)$.

In other words, $\Pi(t,t')$ is a prediction of the function $\varphi$ values on the set $W \times \{t'\}$. These values are not known in the present moment of the time $t$.
To formulate this definition in the case where $t' \to \infty$ we need the existence of the limit $\varphi(w,\infty) = \lim_{t\to\infty} \varphi(w,t)$. In this case, $\Pi(t,t')$ is the prediction of the limit behaviour of the system when $t \to \infty$. Such a dynamic problem will be called a limit recognition problem.
The sufficient condition to formulate a limit recognition problem gives the following simple statement:

**Statement**

the graph $\Gamma$ of allowed movements has no cycles, then for every $w \in W$ the limit exists

$$\varphi(w,\infty) = \lim_{t\to\infty} \varphi(w,t). \tag{4.5}$$

**Proof**

Because the graph $\Gamma$ is finite and has no cycles, the function $\varphi(w,t)$ for each fixed $w$ and $t \to \infty$ may change its values only a finite number of times. Therefore, if $t$ is big enough, the function $\varphi(w,t)$ does not depend of $t$. This proves the statement.

Below, we consider dynamic recognition problems which satisfy the following assumptions (which for example take place for the problem of recognition of strong earthquake prone-areas, see Chapter 6).

1. function $\Phi$ does not depend on time $t$ and is known for the entire set $W$; e. g., $R_\Phi(t) = W \times [T_0, t]$,

2. $R_\varphi(t) = W_0 \times \{t\}$, where $W_0 \subseteq W$. In other words, the prediction $\Pi(t,t')$ is constructed on the basis of just an appearence of the system at the present moment $t$ (this assumption is connected with the construction of applied algorithms).

The subset $W_0 \subseteq W$ is called the learning set and its decomposition

$$P(t){:}\, W_0 = \bigsqcup_{i=1}^{r} \left(W_i(t) \bigcap W_0\right) \tag{4.6}$$

is called the initial decomposition at the moment $t$. It is easely seen that, if the assumptions 1) and 2) are satisfied, then the learning material $\mathcal{O}(t)$ is equal to the initial decomposition $P(t)$ of the learning set $W_0$.

The definition of a dynamic recognition problem establishes a subset of classification problems. On the other hand, we may consider this definition as one which generalizes the classical problem of pattern recognition where the classification of objects $w \in W \setminus W_0$ is unknown, but does not change as time evolves.

In fact, according to [] the classical pattern recognition classification problem is formulated in the following way:

*"Objects of recognition are defined by a set of features, a number of sample objects is known and their descriptions give us the initial (learning) information".*

On the basis of this information an algorithm is synthesized, which makes decisions concerning classification of new coming objects. It is clear that we obtain the same kind of problem if in a dynamic recognition problem the assumptions 1) and 2) are satisfied and the function $\varphi$ does not depend on time. We call such a problem a statistical recognition problem.

There are a wide circle of examples of statistical recognition problems including the exploration of mineral deposits (see [ ]). Let us consider, for example, a problem of recognition of oil bearing layers [ ]. The set of objects in this case is a set of lithological layers in some oil bearing basin. The problem under consideration is to obtain the decomposition $\Pi: W = \mathcal{H} \bigsqcup \mathcal{B}$, where $\mathcal{H}$ is the set of oil bearing and $\mathcal{B}$ the set of water bearing layers. Furthermore, for some subset of layers $W_0 \subset W$ we know the object classifications from the drilling experiments. Thus, we have the initial decomposition $P: W_0 = \mathcal{H}_0 \bigsqcup \mathcal{B}_0$.

In this problem, every object $w \in W$ is either oil bearing or water bearing and the status of the object does not change in time. However, the classification is known only for objects $w \in W_0$. The goal of the prediction is to extend the initial decomposition $P: \mathcal{H}_0 \bigsqcup \mathcal{B}_0$ to a decomposition of the whole set $\Pi: W = \mathcal{H} \bigsqcup \mathcal{B}$.

Therefore, in a statistical problem, $W_0$ is necesseraly satisfied by the condition $W_0 \subset W$, $W_0 \neq W$. Otherwise, the problem is trivial.

In a dynamic problem, it is possible that $W_0 = W$. Furthermore, the assumption $W_0 = W$ makes simplier the formulation of the results below. Thus, below, we shall assume that $W_0 = W$ unless it is said to the contrary.

## 4.2 Stability of Prediction in Dynamic Classification Problems

The classification dependance on time allows us to distinguish a class of predictions, which satisfies an important condition which we call the *stability condition*. We formulate it using the information given in the statement of the problem.
Let us assume that at time $t_0$ we have the decomposition $\mathcal{P}(t_0)$ and that using algorithm $\mathcal{A}$ we obtain a prediction $\Pi(t_0, t')$, $t' \leq \infty$.
Then we consider the same problem at a later time $t_1$ $(t_0 < t_1 < t')$. We call $\mathcal{P}(t_1)$ the permitted decomposition, if the graph $\Gamma$ (see 4.1) permits transfers of objects such that at times $t_0$, $t_1$, and $t'$, the set $\mathcal{W}$ is represented correspondingly by the decompositions $\mathcal{P}(t_0)$, $\mathcal{P}(t_1)$ and $\Pi(t_0, t')$.

**Definition 4.2.1** Prediction $\Pi(t_0, t')$ obtained by algorithm $\mathcal{A}$ is called stable on the time span $(t_0, t')$ if for any $t_1$, such that $t_0 < t_1 < t'$, and for any permitted decomposition $P(t_1)$, considered as initial, the algorithm $\mathcal{A}$, for some values of its free parameters, gives the prediction $\Pi(t_1, t') = \Pi(t_0, t')$.
In other words, stability of prediction means that any change of the learning material in time, not contradictory to the prediction $\Pi(t_0, t')$, does not change the result.
Therefore, unstable predictions in dynamic pattern recognition problems are self contradictory and should not be considered while searching for the precise classification $\Pi(t_0, t')$. On the other hand, stability is not, strictly speaking, a necessary condition of the prediction reliability. The stability of $\Pi(t_0, t')$ follows from the stability of $\Pi(t_1, t')$ $(t_0 < t_1 < t')$, but the opposite is not true. A prediction may stabilize due to the time change.
In connection with applications which are considered below, we shall study a particular case of the limit-pattern recognition problem with decomposition of the set $\mathcal{W}$ into two classes $(r = 2)$. We symbolize this decomposition by $\mathcal{W} = \mathcal{B}(t_0) \bigsqcup \mathcal{H}(t_0)$, where, by definition, $\mathcal{W}_0 = \mathcal{W}$.
We also assume that the graph $\Gamma$ has the appearance $\bullet^2 \longrightarrow \bullet^1$. This permits object transfers from the class $\mathcal{H}(t)$ into the class $\mathcal{B}(t)$ only. Thus $\Gamma$ has no cicles and according to the proposition 4.1 the dynamic pattern recognition problem is correctly stated.
In this case, the initial decomposition has the appearance $\mathcal{P}(t_0)\colon \mathcal{W} = \mathcal{B}_0(t_0) \bigsqcup \mathcal{H}_0(t_0)$, prediction has the appearance $\Pi(t_0) = \Pi(t_0, T_1) : \mathcal{W} = \mathcal{B}(t_0) \bigsqcup \mathcal{H}(t_0)$, $T_1 = \infty$, and all permitted decompositions $\mathcal{P}(t_1), t_0 < t_1 <$

$T_1$ are $\mathcal{P}(t_1) : \mathcal{W} = \mathcal{B}(t_1) \bigsqcup \mathcal{H}(t_1)$, where $\mathcal{B}_0(t_0) \subseteq \mathcal{B}(t_1) \subseteq \mathcal{B}(t_0)$.
We may also describe the permitted decompositions as follows. Let us introduce the set $\mathcal{S} = \mathcal{B}(t_0) \setminus \mathcal{B}_0(t_0)$ of all the objects, which according to the prediction $\Pi(t_0)$ moves from the class $\mathcal{H}_0(t_0)$ to the class $\mathcal{B}(t_0)$. Then any permitted decomposition $\mathcal{P}(t_1)$ has an appearence $\mathcal{P}(\mathcal{S}')\colon \mathcal{W} = (\mathcal{B}_0(t_0) \cup \mathcal{S}') \bigsqcup (\mathcal{H}_0(t_0) \setminus \mathcal{S}')$ for some $\mathcal{S}' \subseteq \mathcal{S}$.
Let the prediction $\Pi(t_0)$ be obtained by an algorithm $\mathcal{A}$ based on an initial decomposition $\mathcal{P}(t_0)$ and on values of the algorithm's free parameters $\mathcal{C}(t_0)$:

$$\mathcal{A}\,[\mathcal{P}(t_0),\ \mathcal{C}(t_0)]\ :\ \mathcal{W} = \mathcal{B} \bigsqcup \mathcal{H}. \tag{4.7}$$

Then the *stability condition* can be defined in the following way:

**Definition 4.2.2** Prediction (4.7) is stable if for every $\mathcal{S}' \subseteq \mathcal{S}$ there exist values $\mathcal{C}(\mathcal{S}')$ of the algorithm $\mathcal{A}$ free parameters, such that the classification

$$\mathcal{A}(\mathcal{P}(\mathcal{S}'), \mathcal{C}(\mathcal{S}')) : \mathcal{W} = \mathcal{B}(\mathcal{S}') \bigsqcup \mathcal{H}(\mathcal{S}')$$

is equal to (3.7).

## 4.3 Stability of Prediction, Algorithm VEF-0

Recall that the algorithm VEF-0 gives the prediction

$$\Pi(t_0) = \Pi(t_0,\ T_1) : \Omega_n = \Lambda_B(t_0) \bigsqcup \Lambda_H(t_0),\ T_1 = \infty,$$

where $\Lambda_B(t_0)$ is a ball with center in the kernel $Z$ and radius $R$. Therefore, according to the definition 4.2.2 the prediction $\Pi(t_0)$ is stable if for any permitted decomposition $\mathcal{P}(t)$, $t_0 < t < T_1$ there is a value of radius $R(\mathcal{P}) = R(\mathcal{P}(t))$ such that the prediction $\Pi(t,\ T_1)$ is equal to $\Pi(t_0, T_1)$; e.g., $\Lambda_B(t) = \Lambda_B(t_0)$. Thus, the following theorem holds:

**Theorem A** For stability of prediction $\Pi(t_0,\ T)$ obtained using the algorithm VEF-0, it is sufficient that for any permitted decomposition $\mathcal{P}$ considered as initial, the kernel $Z(\mathcal{P})$ remains fixed; e.g., $Z(\mathcal{P}) = Z$.
The conditions of the theorem A may be rather simply verified. As is shown in 3.2, the values of the first component of kernel $Z$ depend only on the signum of the determinant

$$\sigma_1' = \det \begin{pmatrix} a_1 & b_1 \\ c_1 & d_1 \end{pmatrix}. \tag{4.8}$$

Furthermore, the values of the matrix (7.8) coefficients depend only on the initial decomposition $\mathcal{P}(t)$. Let $\sigma_l'$, $a_\ell$, $b_\ell$, $c_\ell$, $d_\ell$ correspond to the initial decomposition $\mathcal{P}(t_0) = \mathcal{W} : \mathcal{B}_0(t_0) \bigsqcup \mathcal{H}_0(t_0)$. We symbolize by $m_\ell$ (or $k_\ell$) the number of objects $w \in \mathcal{S}$ such that for $\omega = \Phi(w)$ we have $\omega(\ell) = 0$ (or $\omega(\ell) = 1$). Then minimum and maximum values $\sigma_1'$ are calculated by the formulas :

$$\begin{array}{lll} Q_\ell = \max \sigma_\ell' & = (a_\ell + m_\ell)d_\ell - (b_\ell - m_\ell)c_\ell & = \sigma_\ell' + m_\ell(c_\ell + d_\ell) \\ q_\ell = \min \sigma_\ell' & = a_\ell(d_\ell - k_\ell) - b_\ell(c_\ell + k_\ell) & = \sigma_\ell' - k_\ell(a_\ell + b_\ell) \end{array} \tag{4.9}$$

Maximums and minimums are considered over the whole set of permitted initial decompositions which correspond to all $\mathcal{S}' \subseteq \mathcal{S}$. Thus the following theorem holds:

**Theorem 4.3.1.** Kernel $Z(\mathcal{P})$ remains fixed relative to all permitted initial decompositions $\mathcal{P}$ if and only if $sgnQ_\ell = sgnq_\ell$ for every $\ell = 1, \ldots, n$. (Here we consider that $sgn(0) = +$).
Upon making the substitutions: $a_\ell \mapsto (a_{\ell+1})$; $b_\ell \mapsto (b_{\ell+1})$; $c_\ell \mapsto (c_{\ell+1})$; $d_\ell \mapsto (d_{\ell+1})$, the sufficient conditions of stability can be represented in another appearence, convenient for further applications. Let us fix some $\ell \in \{1, 2, \ldots, n.\}$ and consider the $\ell_{th}$ component of the kernel and the $\ell_{th}$ coordinates of vectors $\omega \in \Omega_n$. We symbolize $b = \mid \mathcal{B}_0(t_0) \mid$, $s = \mid \mathcal{S} \mid$, $c = \mid \mathcal{H}_0(t_0) \setminus \mathcal{S} \mid$. Let $b_0$, $s_0$, $c_0$ ($b_1$, $s_1$, $c_1$) be the numbers of objects $w$ corresponding to $\mathcal{B}_0(t_0)$, $\mathcal{S}$, $\mathcal{H}_0(t_0) \setminus \mathcal{S}$ for which $\omega = \Phi(w)$ satisfies the condition $\omega(1) = 0$ ($\omega(1) = 1$).
It is not complicated to see that $b_0 = a_1$; $b_1 = c_1$; $b_0 + b_1 = b$; $c_0 + c_1 = c$; $s_0 + s_1 = s$.
As follows from the formulas 4.9, the kernel $Z(\mathcal{P})$ for each permitted decomposition $\mathcal{P}(\mathcal{S}') : \mathcal{W} = (\mathcal{B}_0 \bigcup \mathcal{S}') \bigsqcup (\mathcal{H}_0 \setminus \mathcal{S}')$ is defined by the signum of the difference between the following functions of $\mathcal{S}'$:

$$\begin{array}{l} \mathcal{P}_\ell(\mathcal{B}_0 \bigcup \mathcal{S}') = \dfrac{\sum_{w \in \mathcal{B}_0} \omega(\ell) + \sum_{w \in \mathcal{S}'} [\omega(\ell) + 1]}{\mid \mathcal{B}_0 \mid + \mid \mathcal{S}' \mid + 2} \\ \mathcal{P}_\ell(\mathcal{H}_0 \setminus \mathcal{S}') = \dfrac{\sum_{w \in \mathcal{H}_0} \omega(\ell) + \sum_{w \in \mathcal{S}'} [\omega(\ell) + 1]}{\mid \mathcal{H}_0 \mid + \mid \mathcal{S}' \mid + 2} \end{array} \tag{4.10}$$

It is not difficult to prove that maxima and minima of the functions $\mathcal{P}_\ell(\mathcal{B}_0 \bigcup \mathcal{S}')$ and $\mathcal{P}_\ell(\mathcal{H}_0 \setminus \mathcal{S}')$ over the whole set of $\mathcal{S}' \subseteq \mathcal{S}$ have the following appearence:

$$
\begin{aligned}
\mathcal{P}_{\mathcal{B}} = \mathcal{P}^{\ell}_{\mathcal{B}} &= \max_{\mathcal{S}'} \mathcal{P}_{\ell}(\mathcal{B}_0 \bigcup \mathcal{S}') &&= \frac{b_1 + s_1 + 1}{b + s_1 + 2} \\
p_{\mathcal{B}} = p^{\ell}_{\mathcal{B}} &= \min_{\mathcal{S}'} \mathcal{P}_{\ell}(\mathcal{B}_0 \bigcup \mathcal{S}') &&= \frac{b_1 + 1}{b + s_0 + 2} \\
\mathcal{P}_{\mathcal{H}} = \mathcal{P}^{\ell}_{\mathcal{H}} &= \max_{\mathcal{S}'} \mathcal{P}_{\ell}(\mathcal{H}_0 \setminus \mathcal{S}') &&= \frac{c_1 + s_1 + 1}{c + s_1 + 2} \\
p_{\mathcal{H}} = p^{\ell}_{\mathcal{H}} &= \min_{\mathcal{S}'} \mathcal{P}_{\ell}(\mathcal{H}_0 \setminus \mathcal{S}') &&= \frac{c_1 + 1}{c + s_0 + 2}
\end{aligned}
\qquad (4.11)
$$

**Theorem 4.3.2** Kernel $Z(\mathcal{P})$ is fixed for any permitted initial decomposition $\mathcal{P}$ if and only if $sgn(\mathcal{P}_{\mathcal{B}} - p_{\mathcal{H}}) = sgn(p_{\mathcal{B}} - \mathcal{P}_{\mathcal{H}})$ for all $\ell = 1, \ldots, n$.

Theorems 4.3.1 and 4.3.2 give necessary and sufficient conditions for the kernel to remain fixed. At the same time, according to theorem A, the fact that the kernel is fixed is sufficient for the prediction stability. Thus, we may verify stability without consideration of all the subsets $\mathcal{S} \subseteq \mathcal{S}'$ to obtain corresponding classifications. Because the formulas (3.11) are rather simple the verification may be done non-numerically.

The condition of theorem A (i.e., the fact that the kernel $Z(\mathcal{P})$ remains fixed) is not generally speaking necessary for stability of the prediction obtained using algorithm VEF-0. This is illustrated by the following example. Let us consider the space $\Omega_2$ and $\Phi(w) = \{(0,1), (1,0)\}$. Then, for $R = 1$ and the kernels $Z = (0,0)$ and $Z = (1,1)$ the algorithm VEF-0 gives identical classifications.

The condition of theorem A becomes necessary if we consider a narrower circle of the coding maps $\Phi$.

**Theorem B** Let the coding map $\Phi : \mathcal{W} \to \Omega_n$ be a surjection (e.g. $\Phi(\mathcal{W}) = \Omega_n$) and $\mathcal{P} : \mathcal{W} = \mathcal{B}_0 \bigsqcup \mathcal{H}_0$ be an initial decomposition. Then for stability of prediction (VEF-0) $(\mathcal{P}, R) : \mathcal{W} = \mathcal{B} \bigsqcup \mathcal{H}$ obtained for any permitted decomposition $\mathcal{P}(\mathcal{S}')$, considered as initial, it is necessary that the kernel $Z(\mathcal{P})$ remains fixed.

Theorem B is a particular case of a more general theorem which will be proven below (an independent proof is given in [79]).

The following graph illustrates the connections of the obtained theorems for the algorithm VEF-0.

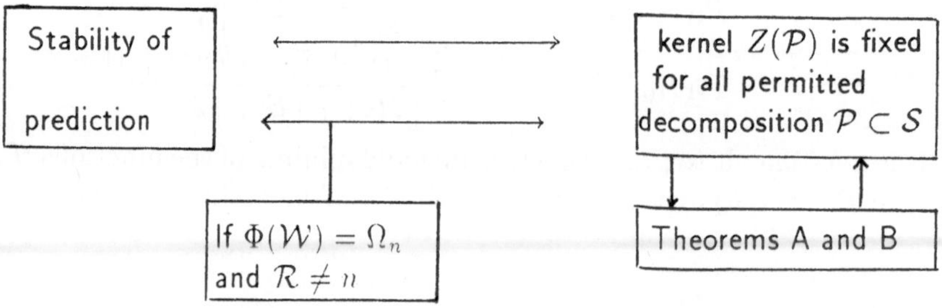

## 4.4 Stability Conditions in the Case of the Algorithm VEF-1

As we know, the algorithm VEF-1 gives the prediction $\Pi(t_0,\ T_1) : \Omega_n = \Lambda_{\mathcal{B}}(t_0) \bigsqcup \Lambda_{\mathcal{H}}(t_0)$, where $\Lambda_{\mathcal{B}}(t_0)$ is a ball of radius $R$ with center in the kernel $Z$ in a metric space $(\Omega_n,\ \rho^\sigma)$. Here $\boldsymbol{\sigma} = (\sigma_1,\ \ldots,\ \sigma_n)$, $\sigma_i \geq 0$ is a vector which does not depend on an initial decomposition $\mathcal{P} = \mathcal{P}(t_0)$. At the same time an arbitrary ball $\mathcal{B}(\omega_0, R) \subset (\Omega_n,\ \rho^\sigma)$ may have more than one center. In other words, it may exist at different points $\omega_1,\ \ldots,\ \omega_n \in \Omega_n$ and have different radii $R_1, \ldots, R_k$ such that $\mathcal{B}(\omega_0, R) = \mathcal{B}(\omega_1, R_1) = \ldots = \mathcal{B}(\omega_k, R_n) = \mathcal{B}$.

**Definition 4.4.1** The set $\{\omega_0,\ \ldots,\ \omega_k\}$ of all the centers of $\mathcal{B}$ is called the centroid of the ball $\mathcal{B}$.

**Example**. A ball of the radius $r$ in $(\Omega_n,\ \rho^\sigma)$ is equal to its centroid. We may make the stability conditions obtained in 4.3 weaker in this case of the algorithm VEF-1.

**Theorem A** For stability of the prediction

$$\Pi(t_0) = \Pi(t_0, T_1) \colon \mathcal{W} = \mathcal{B}(t_0) \bigsqcup \mathcal{H}(t_0),\ T_1 = \infty$$

obtained by the algorithm VEF-1, it is sufficient (and necessary, in the case of a surjective coding $\Phi$) that, for any permitted initial decomposition $\mathcal{P}(\mathcal{S}')$ kernel $Z(\mathcal{P})$ belongs to the centroid of the ball $\Lambda_{\mathcal{B}}(t_0) = \Phi(\mathcal{B}(t_0))$.
To do constructive verification of stability in this case, we need to know the classification of all possible centroids in the space $(\Omega_n,\ \rho^\sigma)$. Let us construct such classification.
Recall that a face of dimension $k$ defined by the edges of the length $\sigma_{i_1}, \ldots, \sigma_{i_k}$ in the metric space $(\Omega_n,\ \rho^\sigma)$ is a subset $\Gamma \subseteq \Omega_n$ which consists of vectors such that the coordinates with numbers $i_1,\ \ldots,\ i_k$ are arbitrarary (e.g. equal to 0 or 1) while the other $(n-k)$ coordinates are fixed.
It is natural to assume that $\sigma_1 \leq \sigma_2 \leq \ldots \leq \sigma_n$. We symbolize also: $\sigma_0 = 0,\ \sigma_{n+1} = \infty$ (4.4.1)

The sought after classification of centroids gives the following

**Theorem B**. Let us consider all $k \in \{0, \ldots, n\}$ such that $\sum_{i=0}^{k} \sigma_i < \sigma_{k+1}$. The set of all centroids in $(\Omega_n, \rho^\sigma)$ is equal to the set of faces defined by the edges of the length $\sigma_1, \ldots, \sigma_k$.

**Remark** Because of (4.4.1.), $\sigma_0$ and $\sigma_{n+1}$, $(k = 0$, and $k = n)$ always satisfy the conditions of the theorem B. In the first case, the centroids are the points (vectors) from $\Omega_n$. In the second case, centroid is the whole set $\Omega_n$. If there is no other $k$ which satisfies the conditions of the theorem B, then all the balls in $(\Omega_n, \rho^\sigma)$ except $\mathcal{B} = \Omega_n$ have only one center. Such a situation holds, for example, in the case $\sigma = (1, \ldots, 1)$, which corresponds to the algorithm VEF-O. Therefore, because of the theorem B, theorem 4.3 B is a particular case of the theorem A.
To proof the theorem B, we need several definitions and lemmas.
Let us fix a point $\omega_0 \in \Omega_n$ and a radius $R$. Following the symmetry of $\Omega_n$ and without limiting the generality, we may assume that $\omega_0 = (0, \ldots, 0)$. We consider an arbitrary set $\alpha = \{i_1, \ldots, i_k\} \subseteq \{1, \ldots, n\}$ and associate with $\alpha$ a face $\Gamma_\alpha$ which contains the point $\omega_0$ and is defined by the edges of length $\sigma_{i_1}, \ldots, \sigma_{i_k}$. By definition, $\Gamma_\alpha$ consists of the vectors $\omega \in \Omega_n$ such that the coordinate numbers $i_1, \ldots, i_k$ are arbitrarary while the other coordinates are zero.

**Definition 4.4.2** A subset $\alpha$ we will be called an $\mathcal{R}$-selection, if $\sigma_{i_1} + \ldots + \sigma_{i_k} \leq \mathcal{R}$. An $\mathcal{R}$-selection $\alpha$ is called complete, if for any $j \in \{1, \ldots, n\} \setminus \alpha$, $\sigma_{i_1} + \ldots + \sigma_{i_k} + \sigma_j > \mathcal{R}$.
It is obvious that any $\mathcal{R}$-selection may be extended to a complete one.

**Definition 4.4.3** For given $\mathcal{R}$, let $\{\alpha_1, \ldots, \alpha_\ell\}$ be the set of all complete $\mathcal{R}$-selections. The intersection

$$\alpha_0 = \bigcap_{i=1}^{\ell} \alpha_i$$

is called an $\mathcal{R}$-kernel.

**Lemma 4.4.1** *Ball* $\mathcal{B}(\omega_0, \mathcal{R})$ *is the union of all faces* $\Gamma_{\alpha_1}, \ldots, \Gamma_{\alpha_\ell}$, *which corresponds to all* $\mathcal{R}$*-selections* $\alpha_1, \ldots, \alpha_\ell$. *The centroid of the ball* $\mathcal{B}(\omega_0, R)$ *is the intersection* $\Gamma_{\alpha_1}, \ldots, \Gamma_{\alpha_\ell}$ *which is the face* $\Gamma_{\alpha_0}$, *where* $\alpha_0$ *is the* $\mathcal{R}$*-kernel.*

**Proof:** For all $\alpha_i$ we have that $\Gamma_{\alpha_i} \subseteq \mathcal{B}(\omega_0, \mathcal{R})$. In fact, for each $\omega \in \Gamma_{\alpha_i}$,

$$\rho(\omega, \omega_0) = \sum_{j=1}^{k} \sigma_{i_j} \mid \omega(i_j) - \omega_0(i_j) \mid \leq \sum_{j=1}^{k} \sigma_{i_j} \leq \mathcal{R} \tag{4.12}$$

Now let $\omega \in \mathcal{B}(\omega_0, \mathcal{R})$. We shall show that $\omega \in \Gamma_{\alpha_i}$ for some $i$. We introduce the selection $\alpha_\omega = \{i_1, \ldots, i_s\}$, which consists of all indices $i_j$, for which $\omega(i_j) = 1$. The selection $\alpha_\omega$ is an $\mathcal{R}$-selection, because $\sigma_{i_1} + \ldots + \sigma_{i_s} = \rho^\sigma(\omega_0, \omega) \leq \mathcal{R}$. We extend $\alpha_\omega$ up to the complete selection $\alpha_i$. Then it is simple to see that $\omega \in \Gamma_{\alpha_i}$. Therefore, we have proven that the ball $\mathcal{B}(\omega_0, \mathcal{R})$ coincides with the unification of faces that correspond to all complete $\mathcal{R}$-selections.

Let $\alpha_0$ be an $\mathcal{R}$-kernel. Then any point from $\Gamma_{\alpha_0}$ is a center of the ball $\mathcal{B}(\omega_0, \mathcal{R})$. In fact, let $\omega^\star \in \Gamma_{\alpha_0}$. Then we consider the mapping $\mathcal{F} : \omega' \mapsto \omega''$ which is the mapping $\Omega_n \to \Omega_n$ such that $\omega'(i) = \omega''(i)$, if $\omega^\star(i) = 0$ while $\omega'(i) = 1 - \omega''(i)$, if $\omega^\star(i) = 1$. It is not complicated to see that $\mathcal{F}$ is an isometric mapping (the distance remains the same). Furthermore, $\mathcal{F}(\omega_0) = \omega^\star$, $\mathcal{F}(\Gamma_{\alpha_0}) = \gamma_{\alpha_0}$ and $\mathcal{F}(\Gamma_{\alpha_i}) = \Gamma_{\alpha_i}$, $i = 1, \ldots, \ell$. Thus, if $\omega_0$ is a center of the ball under consideration, then any point $\omega^\star \in \Gamma_{\alpha_0}$ is also a center.

On the other hand a point $\omega^\star$ which does not belong to $\Gamma_{\alpha_0}$ can not be a center of the ball $\mathcal{B}(\omega_0, \mathcal{R})$. Really, because $\omega^\star \notin \Gamma_{\alpha_0}$, then there exists an $\alpha_i$ such that $\omega^\star \notin \Gamma_{\alpha_i}$. But then the condition $\rho^\sigma(\omega^\star, \omega) \leq \mathcal{R}$ for all $\omega \in \Gamma_{\alpha_i}$ is contradictory to the fact that $\mathcal{R}$-selection $\alpha_i$ is complete. The lemma is proven.

**Consequence** Every centroid in $(\Omega_n, \rho^\sigma)$ is a face.
Lemma 4.4.1 is illustrated by the following

**Example:** Let $n = 3$, $\boldsymbol{\sigma} = (\sigma_1, \sigma_2, \sigma_3) = (1, 2, 2)$. If $\mathcal{R} = 3$ then there are only two complete $\mathcal{R}$-selections: $\alpha_1 = \{1, 2\}$ and $\alpha_2 = \{1, 3\}$. Therefore the $\mathcal{R}$-kernel is $\alpha_0 = \alpha_1 \cap \alpha_2 = \{1\}$. The ball $\mathcal{B}\{\omega_0, \mathcal{R}\} = \Gamma_{\alpha_1} \cup \Gamma_{\alpha_2}$ consists of six points and its centroid consists of two points.

If $\mathcal{R} = 4$ in this space there are three complete $\mathcal{R}$-selections: $\alpha_1 = \{1, 2\}$, $\alpha_2 = \{1, 3\}$, $\alpha_3 = \{2, 3\}$. The $\mathcal{R}$-kernel in this case is the empty set $\emptyset$. The ball $\mathcal{B}(\omega_0, \mathcal{R}) = \Gamma_{\alpha_1} \cup \Gamma_{\alpha_2} \cup \Gamma_{\alpha_3}$ contains seven points. However its centroid contains only one point $\omega_0$.

Lemma 4.4.1 transforms the geometrical problem of centroid set description to a combinatoric problem of description of all the $\mathcal{R}$-kernels corresponding to a given vector $\boldsymbol{\sigma} = (\sigma_1, \ldots, \sigma_n)$.

**Lemma 4.4.2** A set $\alpha \subseteq \{1, \ldots, n\}$ is an $\mathcal{R}$-kernel for some $\mathcal{R}$ if and only if, $\alpha = \{1, \ldots, k\}$ and $\sigma_1 + \sigma_2 + \ldots + \sigma_k < \sigma_{k+1}$.

**Proof** *Sufficiency.* Let $\alpha_0 = \{1, \ldots, k\}$ and $\sigma_1 + \sigma_2 + \ldots + \sigma_k < \sigma_{k+1}$. Let us choose any $\mathcal{R}$ such as $\sigma_1 + \ldots + \sigma_k \leq \mathcal{R} < \sigma_{k+1}$. It is not complicated to see that for such $\mathcal{R}$, $\alpha = \{1, \ldots, k\}$ will be the only one complete $\mathcal{R}$-selection, and therefore an $\mathcal{R}$-kernel. As follows from Lemma 4.4.1, this means that the face $\Gamma_{\alpha_0}$, which contains $\omega_0$ and is defined by edges of length $\sigma_1, \ldots, \sigma_k$, is a ball of radius $\mathcal{R}$ equal to its centroid.

*Necessity.* Let us show that for each $\mathcal{R}$, the $\mathcal{R}$-kernel $\alpha_0$ necessarily has the appearance $\alpha_0 = \{1, \ldots, k\}$. Assume the contrary. Let for example $\ell \in \alpha_0, s \notin \alpha_0$ and $s < 1$. Then, there exists a complete $\mathcal{R}$-selection $\beta$, such that $s \notin \beta$ and $\ell \in \beta$. Let us change $\ell$ to $s$ in the selection $\beta$. In other words, we construct the selection $\beta' = (\beta \setminus \ell) \cup s$. Because $\sigma_s \leq \sigma_\ell$, then $\beta'$ is an $\mathcal{R}$-selection ( $\beta$ is an $\mathcal{R}$-selection). It is possible that selection $\beta'$ is not complete, but we cannot add $\ell$, and therefore we cannot add any index $\ell' \geq \ell_0$. Really, in the contrary case, the selection $\beta$ is not complete. Thus the extension $\beta'$ to a complete $\mathcal{R}$-selection may be done only by adding the indices smaller than $\ell$. In the result we get a complete $\mathcal{R}$-selection $\beta''$, which does not contains $\ell$. This is in contradiction to the fact that $\ell$ belongs to an $\mathcal{R}$-kernel. Therefore, every $\mathcal{R}$-kernel has the appearance $\alpha_0 = \{1, \ldots, k\}$.

Thus the only thing which we need to prove is that if $\alpha = \{1, \ldots, k\}$. and $\sigma_1 + \ldots + \sigma_k \geq \sigma_{k+1}$ then $\alpha$ cannot be an $\mathcal{R}$-kernel for any $\mathcal{R}$. Let us assume the contrary. This means that there is a complete $\mathcal{R}$-selection $\beta$ such that $\{1, \ldots, k\} \subseteq \beta$ and $k+1 \notin \beta$. We consider $\beta' = (\beta \setminus \{1, \ldots, k\}) \cup (k+1)$. Because, $\sigma_{k+1} \leq \sigma_1 + \ldots + \sigma_k$, then $\beta'$ is an $\mathcal{R}$-selection, but possibly not a complete one. We extend $\beta'$ to a complete $\mathcal{R}$-selection. We may also add some indices from $\{1, \ldots, k\}$, but not all of them. Otherwise, $\beta$ would not be a complete selection. Thus, we obtain a complete $\mathcal{R}$-selection $\beta''$, which does not contain $\{1, \ldots, k\}$. This is in contrary to the fact that $\alpha = \{1, \ldots, k\}$ is an $\mathcal{R}$-kernel.

The lemma is proven.

Theorem B obviously follows from the lemmas 4.4.1 and 4.4.2.

The fact that all centroids are faces in $\Omega_n$ gives us the possibility to perform constructive verification of stability analogous to the one we did in 4.3. Let the algorithm VEF-1 give the prediction

$$\Pi(t_0)\colon \Omega_n = \Lambda_{\mathcal{B}}(t_0) \bigsqcup \Lambda_{\mathcal{H}}(t_0)$$

where $\Lambda_{\mathcal{B}}(t_0)$ is a ball in $(\Omega_n,\ \rho^{\sigma})$ the centroid of which is the face defined by the edges of length $\sigma_1,\ \ldots,\ \sigma_k$.

**Statement 4.4.1** Kernel $Z(\mathcal{P})$ belongs to the centroid of the ball $\Lambda_{\mathcal{B}}(t_0)$ for any permitted initial decomposition $\mathcal{P}$ when, and only when, the numbers $Q_\ell$ and $q_\ell$ defined by the formulas (4.9) have the same signums for all $\ell = k+1,\ \ldots,\ n$.

**Proof**: If $Q_\ell$ and $q_\ell$ have the same signums for $\ell = k+1,\ \ldots,\ n$ then the components $z(k+1),\ \ldots,\ z(n)$ remain fixed for all permitted initial decompositions $\mathcal{P}(\mathcal{S})$. The components $z(1),\ \ldots,\ z(k)$, however, may change because the theorem B does not influence the inclusion of the kernel $Z(\mathcal{P})$ to the considered centroid. The reverse statement is obvious, which proves the whole statement.

## 4.5 Stability Zone for Prediction Obtained by the Algorithm VEF-2

According to 3.3, based on the initial decomposition $\mathcal{P}(t_0) : \mathcal{W} = \mathcal{B}_0(t_0) \bigsqcup \mathcal{H}_0(t_0)$ the algorithm VEF-2 calculates a vector of weight $\sigma^0 = \sigma(\mathcal{P}(t_0))$, a kernel $Z = Z(\sigma^0)$ and constructs the prediction $\Pi(t_0) = \Pi(t_0, T_1) : \Omega_n = \Lambda_{\mathcal{B}}(t_0) \bigsqcup \Lambda_{\mathcal{H}}(t_0)$. In this case, $\Lambda_{\mathcal{B}}(t_0)$ is a ball in the metric $\rho^{\sigma^0_+}$, with the center in the kernel $Z(\sigma^0)$ where $\sigma^0_+ = (|\, \sigma_1 \,|,\ \ldots,\ |\, \sigma_n \,|)$ and with radius $\mathcal{R}$ of the ball a free parameter of the algorithm. Due to the time change, the initial decomposition $\mathcal{P}(t)$ also changes. Therefore, the kernel $Z(\mathcal{P})$, vector $\boldsymbol{\sigma}(\mathcal{P})$ and metric $\rho^{\sigma+}$ also change. Thus, the stability verification of the algorithm VEF-2 may be divided into two problems.

1. To determine a zone $\mathcal{L} \subseteq \Re^n$ of all the values of the real vector $\boldsymbol{\sigma}$ such that the given set $\Omega_{\mathcal{B}}(t_0)$ is a ball with the center $Z(\sigma)$ in metric $\rho^{\sigma+}$. $\mathcal{L}$ is called the stability zone of prediction $\Pi(t_0)$.

2. To verify that for all permitted initial decompositions $\mathcal{P}$, vector $\boldsymbol{\sigma}(\mathcal{P})$ belongs to the zone $\mathcal{L}$.

In this paragraph we give the solution of the first problem. The second will be solved in 4.7 in a more general situation.
It is natural to assume that all the coordinates of $\sigma^0$ are non-negative. Therefore, $\sigma^0 = \sigma^0_+$ and $z(\sigma^0) = \overline{0}$ (see the symbolization defined in 3.5). This may be achieved by the following transformation of coordinates in $\Omega_n$ : $\omega(\ell) \to 1-\omega(\ell)$ for all $\ell$ such that $\sigma_\ell > 0$. We have that $\Lambda_{\mathcal{B}}(t_0)$ is a ball of radius $R_0$ with center at the point $z(\sigma^0) = \overline{0}$. As follows from Lemma 3.4, the ball $\Lambda_{\mathcal{B}}(t_0)$ is a union of the faces $\Gamma_{\alpha_\ell}$ for all complete $R_0$–selections $\mathcal{A} = \{\alpha_1, \ldots, \alpha_k\}$.
The centroid of this ball is equal to $\Gamma_{\alpha_0}$, where $\alpha_0 = \bigcap_{i=1}^k \alpha_i$ is $R$–kernel. For any move of the vector $\boldsymbol{\sigma}$ inside the stability zone $\mathcal{L}$ the metrics $\rho^\sigma$ deforms, but $\Lambda_{\mathcal{B}}(t_0)$ remains a ball. This is equivalent to the fact that $\mathcal{A} = \{\alpha_1, \ldots, \alpha_k\}$ is still a set of all complete $\mathcal{R}$–selections and $\alpha_0$ is again an $\mathcal{R}$–kernel (for some $\mathcal{R}$, possibly different from $\mathcal{R}_0$).

**Proposition 4.5.1.** The stability zone $\mathcal{L}$ is:

1. symmetrical regarding any hyperplane $\sigma_i = 0$ if $i \in \alpha_0$.
2. on one side of any hyperplane $\sigma_i = 0$ if $i \overline{\in} \alpha_0$. Furthermore, $\mathcal{L}$ is characterized by the condition $\sigma_i > 0$.

**Proof**

1. Let $i \in \alpha_0$, $\sigma \in \mathcal{L}$, $\boldsymbol{\sigma} = (\sigma_1, \ldots, \sigma_n)$. We consider the vector $\boldsymbol{\sigma}' = (\sigma'_1, \ldots, \sigma'_n)$ obtained by reflection of $\boldsymbol{\sigma}$ from the hyperplane $\sigma_i = 0, (\sigma'_i = \sigma_i,\ \sigma'_j = \sigma_j,\ \forall j \notin i)$. Because $\sigma_+ = \sigma'_+$, the metric $\rho^{\sigma+}$ does not change as $\boldsymbol{\sigma}$ changes to $\boldsymbol{\sigma}'$ . In this case, the kernel $z$ moves inside the centroid of the ball $\Lambda_{\mathcal{B}}(t_0)$, which is equal to $\Gamma_{\alpha_0}$, because $i \in \alpha_0$. Thus, if $\boldsymbol{\sigma} \in \mathcal{L}$, then $\boldsymbol{\sigma}' \in \mathcal{L}$.
2. Let now $j \overline{\in} \alpha_0$. If $\sigma_j < 0$, then $z(j) = 1$ and the kernel $z(\sigma)$ does not belong to the centroid. This means that $\boldsymbol{\sigma} \overline{\in} \mathcal{L}$ and the proposition (3.5.1) is proofed.

The proposition 4.5.1 shows that to determine the zone $\mathcal{L}$ it is sufficient to determine its intersection $\mathcal{L}'$ with the positive octant $\Re^n_+ = \{\boldsymbol{\sigma} = (\sigma_1, \ldots, \sigma_n) \in \Re^n : \sigma_i \geq 0\}$.
To obtain $\mathcal{L}$ in the formulas which describe $\mathcal{L}'$ we make the substitution $\sigma_i \mapsto |\sigma_i|$ for all $i \in \alpha_0$. Correspondingly, we shall consider only vectors $\boldsymbol{\sigma} \in \Re^n_+$ for which $z(\sigma) = \overline{0}$.

Thus, we search for a zone $\mathcal{L}' \subseteq \Re^n_+$ such that $\boldsymbol{\sigma} \in \mathcal{L}'$ when, and only when, $\Lambda_{\mathcal{B}}(t_0)$ is a ball in the metric $\rho^\sigma$ with center at the point $\overline{0}$. Because the ball $\Lambda_{\mathcal{B}}(t_0)$ is defined by the set of all complete $\mathcal{R}$-selections $\mathcal{A} = \{\alpha_1, \ldots, \alpha_k\}$, we need to find all $\boldsymbol{\sigma} \in \Re^n_+$ for which $\mathcal{A}$ is a set of all complete $\mathcal{R}$-selections for some $\mathcal{R}$. The condition $\forall \alpha \in \mathcal{A} \Rightarrow \alpha$ is an $\mathcal{R}$-selection that may be represented by the following system

$$\sum_{i \in \alpha} \sigma_i \leq R, \qquad \alpha \in \mathcal{A}. \tag{4.13}$$

It is clear that any $\mathcal{R}$-selection $\gamma$ belongs to one of the set $\alpha_1, \ldots, \alpha_k$ (because any $\mathcal{R}$-selection may be extended to a complete one). Thus, $\beta$ is not an $\mathcal{R}$-selection (*e.g.* $\sum_{j \in \beta} \sigma_j > R$) when, and only when, $\beta$ is not included in any $\alpha \in \mathcal{A}$. Let us introduce the set $\mathcal{B}$ of all minimum (according to the relation of inclusion) sets of indices, which do not belong to any $\alpha \in \mathcal{A}$. The zone $\mathcal{L}'$ is defined by the following condition: $\boldsymbol{\sigma} = (\sigma_1, \ldots, \sigma_n) \in \mathcal{L}'$ when, and only when

$$\exists\, R : \sum_{i \in \alpha} \sigma_i \leq R \quad \forall\, \alpha \in \mathcal{A}, \quad \sum_{j \in \beta} \sigma_j > R \quad \forall\, \beta \in \mathcal{B}. \tag{4.14}$$

Remark We may interpret the set $\mathcal{B}$ using the notion of *dual set systems*, which will be introduced below. If we denote $N = \{1, \ldots, n\}$, $\mathcal{A}' = \{N \setminus \alpha_1, \ldots, N \setminus \alpha_k\}$, then system $\mathcal{B}$ is dual to the system $\mathcal{A}'$.

**Proposition 4.5.2** The condition (4.14) is equivalent to the following system:

$$\sum_{j \in \beta} \sigma_j - \sum_{i \in \alpha} \sigma_i > 0 \quad \forall\, \alpha \in \mathcal{A}, \quad \forall\, \beta \in \mathcal{B}. \tag{4.15}$$

**Proof** It is clear that (4.15) follows from (4.14). In reverse (4.14) follows from (4.15) if $R = \max_{\alpha \in \mathcal{A}} T_\alpha$, $T_\alpha = \sum_{i \in \alpha} \sigma_i$.
The system (4.15) may be represented in the appearence $\sum a_i^{(\alpha,\beta)} \sigma_i > 0$ where
$a_i^{(\alpha,\beta)} = 1$ if $i \in \beta \setminus \alpha$, $a_i^{(\alpha,\beta)} = -1$ if $i \in \alpha \setminus \beta$ and $a_i^{(\alpha,\beta)} = 0$ in all other cases.
Following the terminology of linear programming [6.3], a subset $C \subseteq \Re^n$ will be called a cone if for any point $x \in C$ and for any real number $b > 0$ we have $bx \in C$. Therefore, $\mathcal{L}'$ is a cone. As was already mentioned, to obtain from $\mathcal{L}'$ the zone of stability $\mathcal{L}$ we have to substitute $|\sigma_\ell|$ for $\sigma_\ell$ in (4.15) for all $\ell$ which belong to the $\mathcal{R}$-kernel $\alpha_0$. Therefore, $\mathcal{L}$ is defined by the following system:

$$\sum_{j\in\beta}\sigma'_j - \sum_{i\in\alpha}\sigma'_i > 0, \quad \alpha\in\mathcal{A}, \quad \beta\in\mathcal{B}, \quad \sigma'_\ell = \begin{cases} \mid\sigma_\ell\mid, & \ell\in\alpha_0 \\ \sigma_\ell > 0, & \ell\overline{\in}\alpha_0 \end{cases} . \qquad (4.16)$$

**Proposition 4.5.3** The stability zone $\mathcal{L}$ of prediction obtained by the algorithm VEF.2 is an open convex cone defined by the system (4.16).

**Proof** It is obvious that $\mathcal{L}$ is a cone, which is open in the natural topology of $\Re^n$. Let us proof that $\mathcal{L}$ is convex. The system (4.16) has an appearence $\sum a_i^{(\alpha,\beta)}\sigma'_i > 0$, where the coefficients $a_i^{(\alpha,\beta)} = 0$, or $-1$ (but never $+1$). Really, if $a_i^{(\alpha,\beta)} = 1$, then $1 \in \beta\setminus\alpha$. This means that $1\overline{\in}\alpha$ and, therefore, $1\overline{\in}\alpha_0 = \bigcap_{\alpha\in\mathcal{A}}\alpha$.
The condition $\mathcal{A}-\mid\mathcal{B}\mid> 0$ (but not $\mathcal{A}+\mid\mathcal{B}\mid> 0$) is equivalent to the system of two conditions $\mathcal{A}+\mathcal{B} > 0$, $\mathcal{A}-\mathcal{B} > 0$. We make such a substitution in steps, finally obtaining the system $\sum b_\ell^{(\alpha,\beta)}\sigma_\ell > 0$, which is equivalent to (4.16) and has the appearance of (4.15). From this we obtain that $\mathcal{L}$ is an intersection of half spaces and, therefore, is a convex set. This proves the proposition.
The above results are summarized by the following:

**Theorem 4.5.1** For stability of prediction $\Pi$ obtained using the algorithm VEF-2, it is sufficient (and, in the case of a surjective $\Phi$, necessary) that for any permitted initial decomposition $\mathcal{P}$, vector $\boldsymbol{\sigma}$ $(\mathcal{P})$ belongs to stability cone $\mathcal{L}(\Pi)$ which is defined by the system (4.16).
Therefore, the problem of stability assessment for predictions obtained using the algorithm VEF-2 has been reduced to the verification of the inclusion $\sigma(\mathcal{P}) \in \mathcal{L}$ for all permitted decompositions $\mathcal{P}$. The algorithm of this verification will be introduced below for a wider class of VSF-algorithms, which includes VEF-2.
Let us consider an example of a stability cone. Let $\Phi$ be a surjective coding map, $n=3$, $\boldsymbol{\sigma} = (1,2,2), \mathcal{R}_0 = 3$. The set of all complete $\mathcal{R}_0$-selections has an appearance: $\mathcal{A}' = \{(1,2), (1,3)\}$, where $\mathcal{R}_0$ is the kernel $\alpha_0 = \{(1)\}$. Then $\mathcal{A}' = \{(3), (2)\}$, $\mathcal{B} = \{(2,3)\}$.
The condition (4.14), which defines $\mathcal{L}'$ looks like: $\exists\,\mathcal{R} : \sigma_1+\sigma_2 \le \mathcal{R}$, $\sigma_1 + \sigma_3 \le \mathcal{R}$, $\sigma_2+\sigma_3 > \mathcal{R}$.
Eliminating $\mathcal{R}$ we obtain the system (4.15):
$\sigma_2 + \sigma_3 > \sigma_1 + \sigma_2$, $\sigma_2+\sigma_3 > \sigma_1+\sigma_3$ or $\sigma_2 > \sigma_1$, $\sigma_3 > \sigma_1$. Because $\mathcal{R}_0$ is the kernel $\alpha_0 = \{(1)\}$ we substitute $\sigma_1 \mapsto\mid\sigma_1\mid$ and obtain the system: $\sigma_2 >\mid\sigma_1\mid$, $\sigma_3 >\mid\sigma_1\mid$.

This system defines the stability zone, which is a convex cone in $\Re^3$.

## 4.6 Stability Zone for Prediction Obtained by a VSF Algorithm

Here we consider an algorithm VSF[$\mathcal{U}, \boldsymbol{\sigma}$ $(\mathcal{P})$, where $\mathcal{U} = \{r_1, \ldots, r_N\}$ is a fixed set of features]. A prediction which has been obtained by this algorithm is defined by the vector of weights $\boldsymbol{\sigma}$ $(\mathcal{P}) = (\sigma_1, \ldots, \sigma_N)$ and a threshold of voting $\Delta$. This prediction may be represented as follows:

$$\begin{array}{ll} \Pi = \Pi(\sigma, \Delta) : \mathcal{W} = \mathcal{B} \bigsqcup \mathcal{H}, & \text{where} \\ \mathcal{B} = \left\{ w \in \mathcal{W} : \sum_{\Upsilon \in \mathcal{U}: w \in \Phi^{-1}(\Upsilon)} \sigma(\Upsilon) \geq \Delta \right\}; & \mathcal{H} = \mathcal{W} \setminus \mathcal{B}. \end{array} \tag{4.17}$$

**Definition 4.6.1** Let $\Pi' : \mathcal{W} = \mathcal{B} \bigsqcup \mathcal{H}$ be some decomposition of the set $\mathcal{W}$. A subset $\mathcal{L} \subseteq \Re^N$ we call a stability zone for the decomposition $\Pi'$ and symbolize it by $\mathcal{L} = \mathcal{L}(\Pi')$ if the condition $\sigma \in \mathcal{L}$ is equivalent to the following condition:

$$\exists\, \Delta : \Pi(\sigma, \Delta) \; = \; \Pi' .$$

In particular, if the decomposition $\Pi'$ is a prediction, then $\mathcal{L}(\Pi')$ will be called the stability zone for the prediction $\Pi'$.
Let $\Pi_0$ be a prediction obtained using the algorithm VSF($\mathcal{U}, \boldsymbol{\sigma}$ $(\mathcal{P})$) at the present moment of time $t_0$ andwith an initial decomposition $\mathcal{P}(t_0)$, $\Pi_0 = \Pi(\boldsymbol{\sigma}$ $(\mathcal{P}(t_0))$.
As in 4.5, the problem of stability verification may be divided into two parts:

1. to find the stability zone $\mathcal{L}(\Pi_0) \subseteq \Re^N$.

2. to verify that for all permitted initial decompositions $\mathcal{P}$, vector $\boldsymbol{\sigma}(\mathcal{P})$ belongs to $\mathcal{L}(\Pi_0)$.

To consider the first part of the problem, we need the following geometrical interpretation of prediction $\Pi$ obtained using the algorithm VSF($\mathcal{U}, \boldsymbol{\sigma}(\mathcal{P})$). Each object corresponds to $w \in \mathcal{W}$ an$N-$dimensional vector whose coordinates equal 0 or 1. This defines which features from $\mathcal{U} = \{r_1, \ldots, r_N\}$ contain $w$. For this, we consider the mapping $\Psi : \Omega_N \mapsto \Omega_N$, $\Psi(\omega) =$

$X = (X_1, \ldots, X_N)$, $X_i = 1$, if $\omega \in \Upsilon_i$; $X_i = 0$, if $\omega \overline{\in} r_i$, $i = 1, \ldots, N$. Therefore, we obtain the mappings $w \longrightarrow^{\Phi} \omega \longrightarrow^{\Psi} \boldsymbol{X}$ the composition of which we symbolize by $\mathcal{X}$. Then $\boldsymbol{X} = \mathcal{X}(\omega)$ is the $N$–dimensional vector which we need to construct.

It is simple to see that the function $\hat{F}(w)$ defined in 3.4 can be represented as follows:

$$\hat{F}(w) = \sum_{i=1}^{n} X_i \sigma_i = (\sigma, X(w)) . \tag{4.18}$$

Here $(\sigma, X(w))$ is a scalar product in the natural inclusion $\Omega_N \subseteq \Re^N$. The condition $\hat{F} = (\sigma, \mathcal{X}(w)) \geq \Delta$ for fixed $w$ defines a halfspace in $\Re^N$. According to definition (4.6.1), the stability zone $\mathcal{L}(\Pi)$ for the prediction $\Pi : \mathcal{W} = \mathcal{B} \bigsqcup \mathcal{H}$ is defined by the following condition:

$$\exists\, \Delta : (\sigma, X(w)) \geq \Delta, \quad \forall\, w \in \mathcal{B}, \quad (\sigma, \mathcal{X}(w)) < \Delta \quad \forall\, w \in \mathcal{H} . \tag{4.19}$$

Eliminating $\Delta$ in (4.19) we obtain:

$$(\sigma, \mathcal{X}(w') - \mathcal{X}(w'')) > 0 \quad \forall w' \in \mathcal{B} \quad \forall w'' \in \mathcal{H} . \tag{4.20}$$

Thus, the stability zone $\mathcal{L}(\Pi)$ is determined by the system (4.20) which is an intersection of open halfspaces limited by the hyperplanes going through the origin of the coordinates. From this we obtain:

**Proposition 4.6.1** Stability zone $\mathcal{L}(\Pi)$ of the algorithm VSF$(\mathcal{U}, \boldsymbol{\sigma}(\mathcal{P}))$ is an open convex cone in $\Re^n$.
We symbolize by $\mathcal{M}$ the minimum convex cone in $\Re^n$ which contains all the vectors $\mathcal{X}\ (w') - \mathcal{X}\ (w'')$, $w' \in \mathcal{B}$, $w'' \in \mathcal{H}$. The cone $\mathcal{M}$ is the set of all possible linear combinations of the vectors $\mathcal{X}\ (w') - \mathcal{X}\ (w'')$ with non-negative coefficients. The cone dual to $\mathcal{M}$ is defined by the formula:

$$\mathcal{M}^{\star} = \{y : (y, x) \geq 0 \quad \forall x \in \mathcal{M}\} . \tag{4.21}$$

Thus, according to (4.20), the stability zone $\mathcal{L}(\Pi)$ is an interior of the cone $\mathcal{M}^{\star}$.
The results obtained are summarized as follows:

**Theorem 4.6.1** For stability of prediction $\Pi$ obtained by an algorithm VSF$(\mathcal{U}, \boldsymbol{\sigma}(\mathcal{P}))$, it is necessary and sufficient that for any permitted initial

decomposition $\mathcal{P}$, vector $\boldsymbol{\sigma}(\mathcal{P})$ belongs to the stability cone $\mathcal{L}(\Pi)$; *e.g.*, $\boldsymbol{\sigma}(p)$ satisfies the system (4.20).
Theorem 4.6.1. in a sense generalizes Theorem 4.5.1. At the same time, in the theorem 4.5.1 the condition $\boldsymbol{\sigma}(\mathcal{P}) \in \mathcal{L}$ for all permitted decompositions $\mathcal{P}$ was only sufficient for stability and necessary only in the case of a surjective $\Phi$. Theorem 4.6.1 provides the necessary and sufficient conditions of stability for any coding map $\Phi$. It is connected to the fact that the stability cone in theorem 4.6.1 depends on $\Phi$ (*see* 4.6.3).
Because algorithm VEF-2 is a particular case of a VSF algorithm, the cones defined by formulas (3.5...) and (3.6.3) are equal in the case of a surjective coding map $\Phi$. To illustrate this we consider an example.

We recall that VSF$(\mathcal{U}, \boldsymbol{\sigma}(\mathcal{P}))$ = VEF-2 if $N = n, \mathcal{U} = \{r_1, \ldots, r_n\}$, $r(i) = \{\omega : \omega(i) = 0\}$, $i = 1, \ldots, n$, $\sigma = \sigma^{\mathcal{H}}$. Assuming that $\sigma^{\mathcal{H}}(r_i) \geq 0$ we obtain that the prediction $\Pi : \Omega_n = \Lambda_{\mathcal{B}} \bigsqcup \Lambda_{\mathcal{H}}$ is defined by the formula $\Lambda_{\mathcal{B}} = \{\omega : (\sigma^{\mathcal{H}}, \omega) \leq \mathcal{R}\}$ where $\mathcal{R} = \sum_{i=1}^{n} \sigma_i - \Delta$.
Therefore, $\Lambda_{\mathcal{B}}$ is a halfspace in $\Omega_n$. It is simple to see that in this case $\Psi(\omega) = 1 - \omega = \overline{\omega}$, $\mathcal{X} = 1 - \Phi(\omega)$. Therefore in (4.6.3) we may carry out the following transformation:

$$\mathcal{X}(w') - \mathcal{X}(w'') = (1 - \Phi(w')) - (1 - \Phi(w'')) = \Phi(w'') - \Phi(w').$$

Let us consider the example of the cone from 3.5. We have $n = 3$, $\boldsymbol{\sigma}(\mathcal{P}_0 = \boldsymbol{\sigma}^{\,0} = (1, 2, 2)$, $\mathcal{R}_0 = 3$. The subset $\Lambda_{\mathcal{B}}$ is defined by the condition

$$\Lambda_{\mathcal{B}} = \left\{\omega : (\boldsymbol{\sigma}^{\,0}, \omega) = \omega(1) + 2\omega(2) + 2\omega(3) \leq 3\right\}.$$

Therefore,

$$\Lambda_{\mathcal{B}} = \{(0,0,0),\ (1,0,0),\ (0,1,0),\ (0,0,1),\ (1,1,0),\ (1,0,1)\}$$

$$\Lambda_{\mathcal{H}} = \Omega_n \setminus \Lambda_{\mathcal{B}} = \{(0,1,1),\ (1,1,1)\}\,.$$

Then, $\mathcal{B} = \Phi^{-1}(\Lambda_{\mathcal{B}})$, $\mathcal{H} = \Phi^{-1}(\Lambda_{\mathcal{H}})$. Assuming that $\Phi$ is a surjective mapping, we rewrite the condition (4.6.2) as follows:

$$\exists\, \Delta : (\boldsymbol{\sigma}\,, \Psi(\omega)) \geq \Delta \quad \forall \omega \in \Lambda_{\mathcal{B}}, \quad (\boldsymbol{\sigma}\,, \Psi(\omega)) < \Delta \quad \forall \omega \in \Lambda_{\mathcal{H}}\,.$$

Then by writing $\Lambda_{\mathcal{B}}, \Lambda_{\mathcal{H}}, \Psi(\omega) = 1 - \omega$ we obtain:

$$\begin{aligned} &\exists\, \Delta' : 0 < \Delta',\ \sigma_1 < \Delta',\ \sigma_2 < \Delta',\ \sigma_3 < \Delta', \quad \sigma_1 + \sigma_2 < \Delta', \\ &\sigma_1 + \sigma_3 < \Delta', \quad \sigma_2 + \sigma_3 > \Delta', \qquad \sigma_1 + \sigma_2 + \sigma_3 > \Delta', \\ &\text{where } \Delta' = \sigma_1 + \sigma_2 + \sigma_3 - \Delta \quad . \end{aligned} \tag{4.22}$$

Deducting the first six conditions from the last two, we obtain:

$$\begin{aligned} &\sigma_2 + \sigma_3 > 0, \quad \sigma_1 + \sigma_2 + \sigma_3 > 0, \qquad \sigma_2 + \sigma_3 > \sigma_1 \\ &\sigma_2 + \sigma_3 > 0, \quad \sigma_3 > 0,\ \sigma_1 + \sigma_3 > 0 \quad \sigma_2 > 0 \\ &\sigma_1 + \sigma_2 > 0, \quad \sigma_3 > \sigma_1,\ \sigma_3 > 0, \qquad \sigma_2 > \sigma_1,\ \sigma_2 > 0\,. \end{aligned} \tag{4.23}$$

This is equivalent to the system:

$$\sigma_2 > \sigma_1, \quad \sigma_2 > -\sigma_1, \quad \sigma_3 > \sigma_1, \quad \sigma_3 > -\sigma_1\,, \tag{4.24}$$

or to the system:

$$\sigma_2 > \mid \sigma_1 \mid, \qquad \sigma_3 > \mid \sigma_1 \mid\,. \tag{4.25}$$

This last system has already been obtained in 4.5.

## 4.7 Stability of Prediction in Case of Normal Weights

The second part of the verification of the prediction stability obtained by a VSF algorithm (*see* 4.5.) concerns itslf with verification of the condition $\boldsymbol{\sigma}\,(\mathcal{P}) \in \mathcal{L}(\Pi)$ for any permitted initial decomposition $\mathcal{P}$.
In principle, it is possible to realize such a verification by directly constructing the vector $\boldsymbol{\sigma}\,(\mathcal{P})$ for all permitted decompositions $\mathcal{P}$. However, the number of such $\mathcal{P}$ is $2^{\mathcal{S}}$, where $\mathcal{S} = \mathcal{B}(t_0) \setminus \mathcal{B}_0(t_0)$. Thus, for large sets $\mathcal{S}$, this way of verification is practically speaking, impossible. Here, we will obtain a more effective algorithm to verify the condition $\boldsymbol{\sigma}\,(\mathcal{P}) \in \mathcal{L}(\Pi)$ in the case where $\boldsymbol{\sigma}\,(\mathcal{P})$ is a vector of normal weights. The number of operations of this algorithm will increase linearly as $\mid \mathcal{S} \mid$.

To do this, we need a more detailed description of some features of the normal weights $\boldsymbol{\sigma}\,^{\mathcal{H}}(\Upsilon, \mathcal{P})$ (*see* 3.4).

Because we will only be considering the normal weights here, we will drop the term "normal" and the index "$\mathcal{H}$" in this paragraph.
We fix a set of objects $\mathcal{W}$ and a mapping $\Phi : \mathcal{W} \mapsto \Omega_n$. The function $\sigma(\Upsilon, \mathcal{P})$ depends on two decompositions: $\mathcal{W} = \mathcal{B}_0 \bigsqcup \mathcal{H}_0$ and $\mathcal{W} = \Phi^{-1}(r) \bigsqcup \Phi^{-1}(\bar{r})$.
In other words, the weight is a function $\sigma = \sigma(\mathcal{B}_0, \Phi^{-1}(\Upsilon))$ of the subsets $\mathcal{B}_0, \Phi^{-1}(r) \subseteq \mathcal{W}$. We extend the domain of definition of this function by assuming that $\sigma$ is a function defined on all possible pairs of subsets $\mathcal{A}, \mathcal{B} \subseteq \mathcal{W}$ and given by the formula:

$$\sigma(\mathcal{A}, \mathcal{B}) = \det \begin{pmatrix} \mid \mathcal{A} \bigcap \mathcal{B} \mid & \mid \overline{\mathcal{A}} \bigcap \mathcal{B} \mid \\ \mid \mathcal{A} \bigcap \overline{\mathcal{B}} \mid & \mid \overline{\mathcal{A}} \bigcap \overline{\mathcal{B}} \mid \end{pmatrix} \tag{4.26}$$

where $\overline{\mathcal{A}} = \mathcal{W} \setminus \mathcal{A}$.

**Proposition 4.7.1** For every subset $\mathcal{A}, \mathcal{B} \subseteq \mathcal{W}$, we have

$$\sigma(\mathcal{A}, \mathcal{B}) = -\sigma(\overline{\mathcal{A}}, \mathcal{B}), \qquad \sigma(\mathcal{A}, \mathcal{B}) = \sigma(\mathcal{B}, \mathcal{A}). \tag{4.27}$$

**Proof** The substitution $\mathcal{A} \mapsto \overline{\mathcal{A}}$ is a substitution of the columns in the determinate (4.26) which change its signum. The substitution $\mathcal{A} \mapsto \mathcal{B}$ is equal to simultaneously exchanging the lines and columns in the determinate (4.26) which leaves the determinate invariant.
Let us fix one of the derivatives in $\sigma(\mathcal{A}, \mathcal{B})$, for example $\mathcal{B} = \mathcal{B}_0$. We symbolize the obtained function of $\mathcal{A}$ by $\sigma_{\mathcal{B}_0}(\mathcal{A})$. According to [] the function $q$ defined on all subsets of $\mathcal{W}$ is called the charge on $\mathcal{W}$ if it is additive; *e.g.*, $q(\mathcal{A} \bigsqcup \mathcal{B}) = q(\mathcal{A}) + q(\mathcal{B})$, $\mathcal{A}, \mathcal{B} \in \mathcal{W}$.

**Proposition 4.7.2** Function $\sigma_{\mathcal{B}_0}(\mathcal{A})$ is a charge on $\mathcal{W}$ which satisfies the condition $\sigma_{\mathcal{B}_0}(\mathcal{W}) = 0$.

**Proof** Let us check that $\sigma_{\mathcal{B}_0}$ is additive. Let $\mathcal{A}_1, \mathcal{A}_2 \subseteq \mathcal{W}$ and $\mathcal{A}_1 \bigcap \mathcal{A}_2 = \emptyset$. We introduce $a = \mid \mathcal{A}_1 \bigcap \mathcal{B}_0 \mid$, $b = \mid \mathcal{A}_1 \bigcap \overline{\mathcal{B}_0} \mid$, $c = \mid \mathcal{A}_2 \bigcap \mathcal{B}_0 \mid$, $d = \mid \mathcal{A}_2 \bigcap \overline{\mathcal{B}_0} \mid$, $e = \mid (\overline{\mathcal{A}_1 \bigcup \mathcal{A}_2}) \bigcap \mathcal{B}_0 \mid$, $f = \mid (\overline{\mathcal{A}_1 \bigcup \mathcal{A}_2}) \bigcap \overline{\mathcal{B}_0} \mid$.
Then, according to (4.26), $\sigma_{\mathcal{B}_0}(\mathcal{A}_1 = a(d+f) - b(c+e)$; $\sigma_{\mathcal{B}_0}(\mathcal{A}_2 = c(b+f) - d(a+e)$; $\sigma_{\mathcal{B}_0}(\mathcal{A}_1 \bigcup \mathcal{A}_2) = (a+c)f - (b+d)e$. Thus, $\sigma_{\mathcal{B}_0}(\mathcal{A}_1) + \sigma_{\mathcal{B}_0}(\mathcal{A}_2) = \sigma_{\mathcal{B}_0}(\mathcal{A}_1 \bigcup \mathcal{A}_2)$.
The condition $\sigma_{\mathcal{B}_0}(\mathcal{W}) = 0$ follows from (4.26). Thus, the proposition 4.7.2 is proven.

The proposition 4.7.2 allows us to calculate rather simply the function $\sigma(\mathcal{A},\mathcal{B})$. Let $\mathcal{A} = \{\mathcal{W}'\}$ and $\mathcal{B} = \{\mathcal{W}''\}$ be the one element subsets of $\mathcal{W}$.

As it follows from (4.26)

$$\begin{aligned} \sigma(\mathcal{A},\mathcal{B}) &= \det\begin{pmatrix} 1 & 0 \\ 0 & \mid \mathcal{W} \mid -1 \end{pmatrix} = \mid \mathcal{W} \mid -1, \text{ if } w' \neq w'' \\ \sigma(\mathcal{A},\mathcal{B}) &= \det\begin{pmatrix} 0 & 1 \\ 1 & \mid \mathcal{W} \mid -2 \end{pmatrix} = -1, \text{ if } w' = w'' \end{aligned} \tag{4.28}$$

We may rewrite (4.28) as

$$\sigma(\{w'\},\{w''\}) = \delta_{w'}^{w''} \mid \mathcal{W} \mid -1\,, \tag{4.29}$$

where $\delta_{w'}^{w''}$ is Kronecker symbol which equals 1 if $w' = w''$ and equals 0 if $w' \neq w''$.

From (4.29) and the additivity of $\sigma$ we obtain

$$\sigma(\mathcal{A},\mathcal{B}) = \sum_{w'\in\mathcal{A},w''\in\mathcal{B}} \left[\delta_{w'}^{w''} \mid \mathcal{W} \mid -1\right]. \tag{4.30}$$

Let us fix a feature $r \in \mathcal{U}$ and consider its weight $\sigma_r(\mathcal{B}) = \sigma(\Phi^{-1}(r),\mathcal{W})$. Let $\mathcal{B}$ consist of one element $\mathcal{B} = \{w\}$. Then due to (4.29)-(4.30)

$$\sigma(\{w\}) = \begin{cases} \mid \Phi^{-1}(r) \mid, & \text{if} w \in \Phi^{-1}(r) \\ - \mid \Phi-1(r) \mid, & \text{if} w \overline{\in} \Phi^{-1}(r) \end{cases} \tag{4.31}$$

Because according to Proposition 4.7.2 $\sigma$ is an additive function of argument $\mathcal{B}$ we obtain

$$\sigma_\Gamma(\mathcal{B}) = \sum_{w\in\mathcal{B}} \sigma_\Gamma(\{w\}) \qquad \forall\, \mathcal{B} \subseteq \mathcal{W}\,. \tag{4.32}$$

At the moment of time $t_0$ and on the basis of initial decomposition $\mathcal{P}(t_0) : \mathcal{W} = \mathcal{B}_0(t_0) \bigsqcup \mathcal{H}_0(t_0)$, let an algorithm VSF$(\mathcal{U},\sigma(\mathcal{P}))$ give the prediction $\Pi(t_0) : \mathcal{W} = \mathcal{B}(t_0) \bigsqcup \mathcal{H}(t_0)$. According to 4.2 all permitted initial decompositions $\mathcal{P} : \mathcal{W} = \mathcal{B} \bigsqcup \mathcal{H}$ have an appearence $\mathcal{B} = \mathcal{B}_0(t_0) \bigcup \mathcal{S}'$, where $\mathcal{S}'$ is an arbitrarary subset $\mathcal{S}' \subseteq \mathcal{S} = \mathcal{B}(t_0) \setminus \mathcal{B}_0(t_0)$. From this we obtain:

$$\sigma(r,\mathcal{P}) = \sigma_r(\mathcal{B}) = \sigma_r(\mathcal{B}_{\prime}(t_0)) + \sum_{w\in\mathcal{S}} \sigma_r(\{w\})\,. \tag{4.33}$$

Contributions of elements $w \in \mathcal{S}'$ to $\sigma_r(\mathcal{B})$ are independent and may be simply calculated using the formula (4.30).

The formulas (4.30) and (4.33) give an opportunuity to verify the condition $\sigma(\mathcal{P}) \in \mathcal{L}(\Pi(t_0))$ for all permitted decompositions $\mathcal{P}$. According to 4.6, the stability zone $\mathcal{L}(\Pi(t_0))$ is defined by the system $(\sigma, X^\nu) > 0$, $\nu = 1, \ldots, M$ where $X^1, \ldots, X^M$ are some fixed $N$-dimensional vectors and $\boldsymbol{\sigma} = (\sigma_{r_1}(\mathcal{B}), \ldots, \sigma_{r_N}(\mathcal{B}))$ is vector of weights which depends on $\mathcal{B}$. This system is obviously satisfied if $\mathcal{B} = \mathcal{B}_0(t_0)$ because $\sigma(\mathcal{P}(t_0)) \in \mathcal{L}(\Pi(t_0))$. We must verify that it is also satisfied if $\mathcal{B} = \mathcal{B}_0(t_0) \bigcup \mathcal{S}'$ for all $\mathcal{S}' \subseteq \mathcal{S}$. Let us fix a $\nu \in \{1, \ldots, M\}$ and consider the condition $(\sigma, x^\nu) > 0$.
In this case, we find a minimum $m_\nu$ of $T_\nu(\mathcal{B}) = (\sigma(\mathcal{B}), x^\nu)$ where $\mathcal{B} = \mathcal{B}_0(t_0) \bigcup \mathcal{S}'$ and where the minimum is taken over all $\mathcal{S}' \subseteq \mathcal{S}$. Because $\sigma_\Gamma(\mathcal{B})$ are additive functions and $T_\nu(\mathcal{B}) = \sum_{i=1}^{n} (\sigma_{\Gamma_i}(\mathcal{B}), x_i^\nu)$ is a linear combination, the function $T_\nu(\mathcal{B})$ is also an additive function. This allows us to find a subset $\mathcal{S}_m \subseteq \mathcal{S}$ on which $T_\nu$ reaches its minimum *e.g.* to find $\mathcal{S}_m$ such that $T_\nu(\mathcal{B}_0(t_0) \bigcup \mathcal{S}_m) = m_\nu$.
We fix an object $w \in \mathcal{S}$ and calculate the vector $\sigma(\{w\})$ using formula (4.30).
It is obvious that $w \in \mathcal{S}_m$ when and only when $T_\nu(\{w\}) = [\sigma(\{w\}), x^\nu] < 0$. Considering in this way all of the objects $w \in \mathcal{S}$, we construct the sought for subset $\mathcal{S}_m$. Then, putting $\mathcal{S}' = \mathcal{S}_m$ in (4.33), we calculate $m_\nu = T_\nu(\mathcal{B}) = T_\nu(\mathcal{B}_0(t_0) \bigcup \mathcal{S}_m)$. If $m_\nu \leq 0$, then $\sigma_\Gamma(\mathcal{B}) \overline{\in} \mathcal{L}(\Pi(t_0))$ and therefore, the prediction $\Pi(t_0)$ is not stable. If $m_\nu > 0, \forall\ \nu = 1, \ldots, M$, then we have proven the stability of $\Pi(t_0)$.

The proposed algorithm requires the execution of the same procedure for all the elements of the set $\mathcal{S}$ and, therefore, the number of its operations linearly increases with increasing of $\mid \mathcal{S} \mid$. This is the main advantage of the method constructed in this paragraph compared with a direct construction of vectors $\boldsymbol{\sigma}$ $(\mathcal{P})$ for all permitted $\mathcal{P}$ which has operations increasing as $2^{|\mathcal{S}|}$.

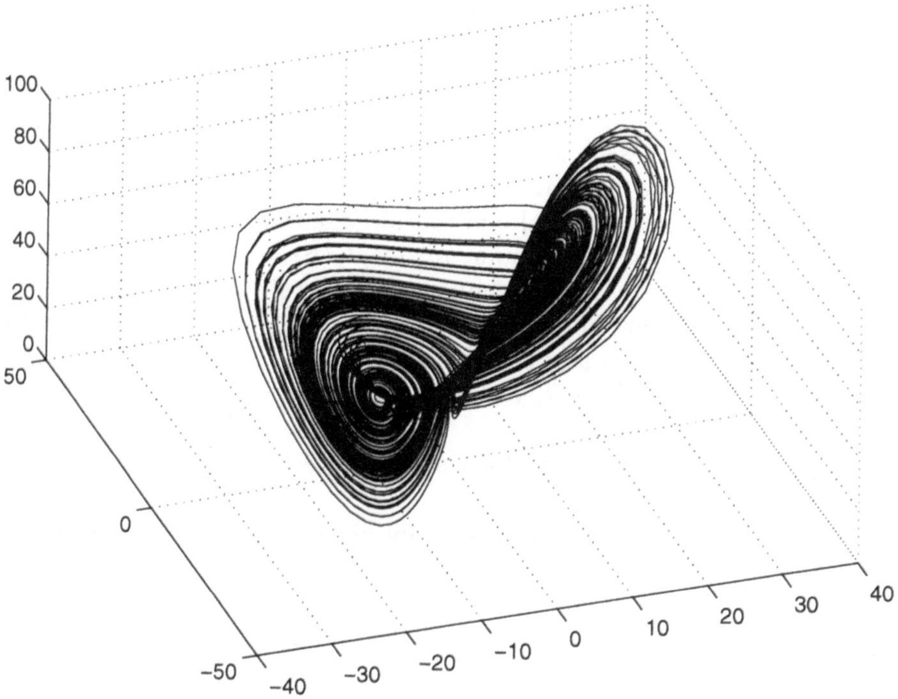


The Lorenz Attractor – The parameters are the same as in page 4, but the perspective representation is different. It shows in the phase space shows 10,000 points discrete phase trajectories (after David AUBERT, 1997).

# Chapter 5

# Dual Systems of Sets and Local Stability of Classification

As it was shown in chapter 3, prediction in dynamic classification problems can be unstable at the moment when we investigate the problem, but can become stable with time. Thus, a natural question arrises : how probable is it that a given unstable prediction will become stabilized in the future? In this chapter we try to answer this question. To do so we use the notion of *dual systems of sets* introduced in [86].

## 5.1 Dual Systems of Sets

Let us fix a finite set $\mathcal{N} = \{a_1, \ldots, a_r\}$. We call by a system of sets (or simply a system) an arbitrary set of subsets from $\mathcal{N}$. Below we will consider two systems: $\mathcal{A} = \{\alpha_1, \ldots, \alpha_k\}$ and $\mathcal{B} = \{\beta_1, \ldots, \beta_\ell\}$.

**Definition 5.1.1** The system $\mathcal{B}$ is called dual to the system $\mathcal{A}$ and symbolized by $\mathcal{A}^\star$ if

1. $\forall\, \beta \in \mathcal{B} \Longrightarrow \beta \cap \alpha \neq \emptyset \quad \forall\, \alpha \in \mathcal{A}$
2. if some set $\gamma \subseteq \mathcal{N}$ such as $\forall\, \alpha \in \mathcal{A}$, $\gamma \cap \alpha \neq 0$ then, $\exists\, \beta \in \mathcal{B}$ such as $\beta \subseteq \gamma$.

It is clear that by definition the dual system $\mathcal{B} = \mathcal{A}^\star$ is not unique. For example, $\mathcal{A}^\star$ can be the system of all subsets $\mathcal{N}$, which overlaps with all $\alpha \in \mathcal{A}$.

We introduce the following equivalence relation on the set of systems:

$\mathcal{A} \sim \mathcal{B}$, if $\forall \alpha \in \mathcal{A} \quad \exists \beta \in \mathcal{B}$ such as $\alpha \subseteq \beta$.

We define the operations of addition and multiplication of the systems by the following formulas:

$$\mathcal{A} \oplus \mathcal{B} = \{\alpha\} \cup \{\beta\}, \quad \mathcal{A} \otimes \mathcal{B} = \{\alpha \cup \beta\} \quad \alpha \in \mathcal{A}, \beta \in \mathcal{B}. \quad (5.1)$$

We also introduce the following relation of inclusion in the set of systems:

$\mathcal{A} \subset \mathcal{B}$ if $\forall\, \alpha \in \mathcal{A} \quad \exists\, \beta \in \mathcal{B}$ such as $\beta \subseteq \alpha$.

It is obvious that $\mathcal{A} \sim \mathcal{B} \iff \mathcal{A} \supset \mathcal{B}$ and $\mathcal{B} \supset \mathcal{A}$.

By the 1-system (unit-system) we mean the system of all subsets of $\mathcal{N}$ and by the 0-system (null system) we mean the empty system.

As is shown in [2.6], the set of the classes of equivalent systems having operations introduced in definition (5.1.1) and in formulas (5.1) and having the above relation of inclusion is a non-complete Boolean algebra.

**Theorem 5.1.1** The non-complete Boolean algebra of classes of equivalent systems, defined in the set $\mathcal{N} = \{a_1, \ldots, a_r\}$ with operations $\oplus$, $\otimes$, $\star$ and relation of inclusion $\mathcal{A} \subseteq \mathcal{B}$ is isomorphic to the non-complete Boolean algebra of all monotonous Boolean functions of $r$ derivatives having operations of disjunction $(\vee)$, conjunction $(\wedge)$, duality $(\star)$ and relations of implication $(F(\mathcal{A}) \to F(\mathcal{B}))$.

**Proof.** The sought for isomorphism is realized by the mapping $\varphi$ that follows. For each element $a_i \in \mathcal{N}$ we construct a Boolean variaty, which is also symbolized by $a_i$.

Furthermore, we correspond to the system $\mathcal{A} = \{\alpha_1, \ldots, \alpha_k\}$ a disjunctive normal form (DNF) $A = \vee_{\alpha \in \mathcal{A}} \wedge_{a \in \alpha} a$.

We symbolize by $F_{\mathcal{A}}$ the corresponding Boolean function. It is obvious that $A$ does not contain rejections and therefore, $F_{\mathcal{A}}$ is a monotonous function. It is clear that $(\mathcal{A} \subset \mathcal{B}) \iff (F_{\mathcal{A}} \to F_{\mathcal{B}})$. Thus, $(\mathcal{A} \sim \mathcal{B}) \iff (\mathcal{A} \subset \mathcal{B}$ and $\mathcal{B} \subset \mathcal{A}) \iff (F_{\mathcal{A}} \to F_{\mathcal{B}}$ and $F_{\mathcal{B}} \to F_{\mathcal{A}}) \iff F_{\mathcal{A}} = F_{\mathcal{B}}$. Therefore, the systems $\mathcal{A}$ and $\mathcal{B}$ are equivalent when and only when they correspond to equal Boolean functions.

Each monotonous function $F$ has a co-image. To find it, it is sufficient to consider the reduced DNF $F$ and the system of sets corresponding to implicants of this form. Therefore, $\varphi$-is a bijective mapping of the set of classes of equivalent systems in the set of monotonous Boolean functions. The 1-system corresponds to the function $F_{\mathcal{E}} \equiv 1$ while the 0-system corresponds to $F_0 \equiv 0$.

It is equivalent that the sum $\mathcal{A}\oplus\mathcal{B}$ corresponds to $F_{\mathcal{A}} \vee F_{\mathcal{B}}$ and the product $\mathcal{A}\otimes\mathcal{B}$ corresponds to $F_{\mathcal{A}} \wedge F_{\mathcal{B}}$. Furthermore, $\mathcal{B} = \mathcal{A}^{\star}$ when, and only when, the functions $F_{\mathcal{A}}$ and $F_{\mathcal{B}}$ are dual.

**Consequence.** If the system $\mathcal{B}$ is dual to system $\mathcal{A}$, then $\mathcal{A}$ is dual to $\mathcal{B}$.

## 5.2 S-Theorems and S-Counter-examples

The technique of dual sets has been introduced in [74] in connection with game theory. As we shall see, it may be useful in more general situations.

**Example.** Let us consider a set $\mathcal{D}$ of objects for which a set of features $N = \{\alpha_1, \cdots, \alpha_r\}$ is defined.
By a feature $\alpha_i$ we mean a mapping $\alpha_i : \mathcal{D} \to [0, 1]$ and by the condition $\alpha_i(d) = 1$ we mean that the object $d \in \mathcal{D}$ possesses a feature $\alpha_i$.
We also introduce a subset $\mathcal{S} \subseteq \mathcal{D}$ of objects $d \in \mathcal{D}$ which have all the features $N$.

**Definition 5.2.1.** A subset $\alpha \subseteq N$ is called an S-theorem, if any object $d \in \mathcal{D}$, which possesses the set of features $\alpha$ belongs to $S$.
A subset $\beta \subseteq N$ is called an S-counter-example, if $\exists d \in \mathcal{D}$ such that $d$ possesses the set of features $\beta$, but does not belong to $S$.
One can see that each subset $\gamma \subseteq N$ is either an S-theorem or an S-counter-example. If $\beta$ is a counter-example and $\beta' \subseteq \beta$, then $\beta'$ is also a counter-example. Also, if $\alpha$ is an S-theorem and $\alpha' \supseteq \alpha$, then $\alpha'$ is also an S-theorem.
The following illustration comes from geometry.
Let $\mathcal{D}$ be the set of convex rectangles on the plane and $\mathcal{S}$ the set of squares. We introduce the following set of features : $\mathcal{N} = \{a_1, a_2, a_3, a_4, a_5\}$ where $a_1$ means the diagonals are orthogonal, $a_2$ means the diagonals are equal, $a_3$ means that among the sides there is a pair of parallels, $a_4$ means that among the angles there is one equal to $90°$ and $a_5$ means that among the sides there is a pair of equal lengths. An example of an S-theorem here is the set $\mathcal{A} = \{a_1, a_2, a_3, a_4\}$ while the set $\mathcal{B} = \{a_1, a_2, a_3, a_5\}$ is an S-counter-example.
We symbolize by $\mathcal{A}$ the system of all S-theorems and by $\mathcal{B}$ the system of

all sets $\mathcal{N} \setminus \beta$ where $\beta$ is an S-counter-example.

**Theorem 5.2.1.** For all $\mathcal{D}$, $\mathcal{S}$ and $\mathcal{N}$, the systems $\mathcal{A}$ and $\mathcal{B}$ are dual.

**Proof** For any S-theorem $\alpha$ and any S-counter-example $\beta$ we have that $\alpha \cap (\mathcal{N} \setminus \beta) \neq \emptyset$. (If not, then $\alpha \subseteq \beta$ which is contradictory to the definition of $\mathcal{A}$).
Let us consider a subset $\gamma \subseteq \mathcal{N}$, such that $(\mathcal{N} \setminus \gamma) \cap \alpha \neq \emptyset$ for all S-theorems $\alpha$. We shall show in this case that there is an S-counter-example $\beta$ such that $\mathcal{N} \setminus \beta \subseteq \mathcal{N} \setminus \gamma \iff \gamma \subseteq \beta$.
Indeed, by definition $\gamma$ does not contain an S-theorem and thus $\gamma$ is an S-counter-example. Therefore, there is a S-counter-example $\beta$ such that $\gamma \subseteq \beta$. This proves the theorem.

Theorem 5.2.1. shows that when searching for S-counter-examples we may complete the search when the obtained systems $\mathcal{A}' = \{\alpha\}$ and $\mathcal{B}' = \{\beta\}$ become dual.
In fact, if we add to $\mathcal{A}'$ and $\mathcal{B}'$ several new S-theorems and S-counter-examples, we will obtain the systems $\mathcal{A}''$ and $\mathcal{B}''$. At the same time, $\mathcal{A}'' \supset \mathcal{A}', \mathcal{B}'' \supset \mathcal{B}'$ and $\mathcal{B}' = (\mathcal{A}')^*$.
Therefore, $\mathcal{B}'' = (\mathcal{A}'')^*$ and $\mathcal{A}'' \sim \mathcal{A}'$, $\mathcal{B}'' \sim \mathcal{B}'$; *e.g.*, for every S-theorem $\alpha'' \in \mathcal{A}''$ there is an $\alpha' \in \mathcal{A}'$ such that $\alpha' \subseteq \alpha''$ and for every new S-counter-example $\beta''$ $(\mathcal{N} \setminus \beta'' \subseteq \mathcal{B}'')$ there is a $\beta'$ $(\mathcal{N} \setminus \beta' \in \mathcal{B}')$ such that $\beta' \supseteq \beta''$.
Thus, if the systems $\mathcal{A}'$ and $\mathcal{B}'$ are dual, then we can add only weaker S-theorems and S-counter-examples.

Coming back to our example with rectangles and squares, we see that the sets

$$\{a_1, a_2, a_3, a_4\}, \quad \{a_1, a_2, a_4, a_5\}, \quad \{a_1, a_3, a_4, a_5\}$$

are S-theorems, and the sets

$$\{a_1, a_2, a_3, a_5\}, \quad \{a_2, a_3, a_4, a_5\}, \quad \{a_1, a_2, a_4\}, \quad \{a_1, a_3, a_4\}, \quad \{a_1, a_4, a_5\}$$

are S-counter-examples.
Let us consider the following systems of sets:

$$\begin{aligned} \mathcal{A}' &= \{(a_1, a_2, a_3, a_4), \quad (a_1, a_2, a_4, a_5)\} \\ \mathcal{B}' &= \{(a_4), (a_1), (a_3, a_5), (a_2, a_5), (a_2, a_3)\} \end{aligned} \tag{5.2}$$

Using the functions $F_{\mathcal{A}'}$ and $F_{\mathcal{B}'}$, it is simple to show that $\mathcal{B}' = (\mathcal{A}')^\star$. Thus the system of all S-theorems $\mathcal{A}$ is equivalent to $\mathcal{A}'$ and the system of the complementary sets to all S-counter-examples is equivalent to $\mathcal{B}'$.

## 5.3 Local Stability and Stabilizing Sets

In section 4.4 we have introduced the stability condition which may be used to evaluate the reliability of a prediction obtained using a classification algorithm with learning. However, the prediction can be unstable at the moment when it has been obtained, but then stabilizes later on.

We now address the primary question of this chapter: how probable is it that a given unstable prediction will stabilize in the future?

To answer this, we introduce the notions of local stability and stabilizing sets. As will be shown, a prediction may stabilize when and only when it is locally stable. If, at the same time, the number of stabilizing sets is large enough, and the number of elements in these sets are small, then the probability of prediction becoming stable is large. On the other hand, if the condition of local stability does not take place then there are no stabilizing sets. Thus, the absence of local stability goes against prediction reliability.

When do we actually have to deal with an unstable prediction $\mathbf{\Pi}$? On the one hand, with time we can have a new initial decomposition from which the considered prediction can become stable. On the other hand, with time we may get a new initial decomposition from which it is not possible to obtain the prediction $\mathbf{\Pi}$ using the given algorithm. Thus, we may explore two possible approaches to evaluate reliability of unstable predictions.

The first approach is to substitute the unstable prediction $\mathbf{\Pi}$ by a weaker, but stable prediction. Let a prediction $\mathbf{\Pi} : \mathcal{W} = \mathcal{B} \bigsqcup \mathcal{H}$, obtained using an algorithm $\mathcal{A}$ be unstable. We distinguish a subset $\mathcal{W}^* \subseteq \mathcal{W}$ of objects that the prediction $\mathbf{\Pi}$ is not able to classify with certaincy. It may happen that the prediction $\mathbf{\Pi}'$, obtained using the same algorithm $\mathcal{A}$ with the same values for the free parameters but using the set of objects $\mathcal{W}' = \mathcal{W} \setminus \mathcal{W}^*$ turns out to be stable.

The second approach is the following. We leave the prediction as it is, but make weaker the stability condition. Let us consider an initial decomposition $\mathcal{P}_0 : \mathcal{B}_0 \bigsqcup \mathcal{H}_0$ and prediction $\mathbf{\Pi}_0 : \mathcal{W} = \mathcal{B} \bigsqcup \mathcal{H}$ which, generally speaking, is not stable for $\mathcal{P}_0$. We recall that $\mathcal{S} = \mathcal{B} \setminus \mathcal{B}_0$.

**Definition 5.3.1.** A subset $\alpha \subseteq \mathcal{S}$ will be called a stabilizing set, if the prediction $\mathbf{\Pi}_0$ is stable for an initial decomposition $\mathcal{P}_\alpha : \mathcal{W} = (\mathcal{B}_0 \bigcup \alpha) \bigsqcup (\mathcal{H}_0 \setminus \alpha)$. The set $\mathcal{X}$ of all subsets from $\mathcal{S}$ which stabilize $\mathbf{\Pi}_0$ will be called the **system of stabilizing sets.**

The system $\mathcal{X}$ depends on the prediction $\mathbf{\Pi}_0$ and the initial decomposition $\mathcal{P}_0$. In other words, $\mathcal{X} = \mathcal{X}(\mathbf{\Pi}_0, \mathcal{P}_0)$. As it is simple to see, the prediction $\mathbf{\Pi}_0$ is stable when and only when $\mathcal{X}$ is equal to the system $2^{\mathcal{S}}$ of all the subsets of $\mathcal{X}$, in other words, when $\alpha = \emptyset$.

**Definition 5.3.2.** We call the prediction $\mathbf{\Pi}_0$ locally stable if $\alpha = \mathcal{S}$ belongs to $\mathcal{X}$.

A lack of local stability means that the prediction $\mathbf{\Pi}_0$ cannot be stabilized with time anyhow. On the other hand, it is not complicated to see that the prediction $\mathbf{\Pi}_0$ is locally stable when and only when it may be obtained from the initial decomposition $\mathcal{P} = \mathbf{\Pi}_0 : \mathcal{W} = \mathcal{B} \bigsqcup \mathcal{H}$. Thus, existence of local stability for prediction $\Pi_0$ does not depend on the initial decomposition $\mathcal{P}_0$.

**Proposition 5.3.1.** Let $\alpha$ be a stabilizing set, which is included in a set $\alpha'$ such that $\alpha \subseteq \alpha' \subseteq \mathcal{S}$. Then $\alpha'$ is also a stabilizing set.

The proof of this proposition becomes obvious if we take into consideration the definition of stability (see ... ).

A system of sets which satisfies the conditions of the proposition 5.3.1. we call as a **monotonous system.**
Let us also consider a system $\mathcal{Y}$ of all subsets $\beta \subseteq \mathcal{S}$ such that prediction $\mathbf{\Pi}_0$ is stable for the initial decomposition $\mathcal{P}_\beta : \mathcal{W} = (\mathcal{B}_0 \bigcup (\mathcal{S} \setminus \beta)) \bigsqcup (\mathcal{H} \setminus \beta)$. The sets $\alpha \in \mathcal{X}$, $\beta \in \mathcal{Y}$ are characterized by the following conditions: if all objects from $\alpha$ are classified as $\mathcal{B}_0$, then it stabilizes the initial prediction $\mathbf{\Pi}_0$, while if all objects from $\mathcal{S}$, excluding those which belongs to $\beta$, are classified as $\mathcal{B}_0$, then this is not sufficient to stabilize the prediction $\mathbf{\Pi}_0$.

**Proposition 5.3.2.**

1. Systems $\mathcal{X}$ and $\mathcal{Y}$ are dual.

2. System $\mathcal{Y}$ is monotonous.

We shall prove more general proposition, from which proposition 5.3.2. will follow.

Let $\mathcal{S} = \{a_1, \cdots, a_r\}$ be a finite set. We consider a system where $\mathcal{X}$ is a system of subsets from $\mathcal{S}$, $\mathcal{E}$-is a system of all subsets $\mathcal{S}$, $\mathcal{Y} = \mathcal{E} \setminus \mathcal{X}$, and $\mathcal{Y} = \{(\mathcal{S} \setminus \mathcal{B}'), \beta' \in \mathcal{Y}'\}$.

**Theorem 5.3.1.** Systems $\mathcal{X}$ and $\mathcal{Y}$ are dual. $\mathcal{X}$ is monotonous when and only when $\mathcal{Y}$ is monotonous.

**Proof.** We correspond to each element $a \in \mathcal{S}$ a Boolean variaty, which we also symbolize by $a$. We make a one to one correspondance between systems of subsets $\mathcal{S}$ and Boolean functions of $a_1, \cdots, a_r$. In this mapping, system $\mathcal{X}$ corresponds to the function $F_{\mathcal{X}} = F(a_1, \cdots, a_r)$ which is defined by the condition $F_{\mathcal{X}} = 1 \iff (\exists \alpha \in \mathcal{X} : a = 1 \iff a \in \alpha)$.
Systems $\mathcal{E}$, $\mathcal{Y}'$, $\mathcal{Y}$ correspond to the following functions : $F_{\mathcal{E}} \equiv 1$, $F_{\mathcal{Y}'} = \overline{F_{\mathcal{X}}}$ (reverse function), and $F_{\mathcal{Y}} = F^{\star}_{\mathcal{X}}$ (dual function).
This follows from definitions of unit, reverse and dual functions. Thus, the functions $F_{\mathcal{X}}$ and $F_{\mathcal{Y}}$ are dual, which means that they are monotonous (or not) simultaneously. From this, if we take into account theorem 4.1.1., we obtain that systems $\mathcal{X}$ and $\mathcal{Y}$ are dual. Finally, it is not difficult to verify that monotonous Boolean functions correspond to monotonous systems and vice versa. Thus theorem 5.3.1. is proven.

The system of stabilizing sets $\mathcal{X}$ characterizes the stability of prediction $\mathbf{\Pi}_0$. It is possible to also show [70, 79, 86] that the relation of implication on the set of systems introduces an order to the sets of predictions according to the stability level. The case $F_{\mathcal{X}} = 1$ corresponds to stable prediction and the case $F_{\mathcal{X}} \equiv 0$ to an absence of even local stability.
We next introduce a method of constructing a system of stabilizing sets for the class of algorithms VEF with normal weights and for the algorithms VEF-0, VEF-1. We will use the following example of dual systems.
Let $\mathcal{S} = \{a_1, \cdots, a_r\}$ and let $f : \mathcal{S} \longrightarrow \Re^{+}$ be a real non-negative function over $\mathcal{S}$. For any $\gamma \subseteq \mathcal{S}$, we denote $f(\gamma) = \sum_{a \in \gamma} f(a)$. Let $f(\mathcal{S}) = C = C_1 + C_2$ where $C, C_1, C_2 > 0$.
We consider the following two systems of subsets from $\mathcal{S}$:

$$\begin{aligned} \mathcal{X} &= \{a : a \subseteq \mathcal{S}, \ f(a) \geq C_1\}, \\ \mathcal{Y} &= \{\beta : \beta \subseteq \mathcal{S}, \ f(\beta) > C_2\}. \end{aligned} \tag{5.3}$$

**Proposition 5.3.3.** Systems $\mathcal{X}$ and $\mathcal{Y}$, defined by (4.3), are monotonous and dual.

**Proof.** Systems $\mathcal{X}$ and $\mathcal{Y}$ are monotonous because the function $f$ is positive. To prove that the systems are dual we apply the theorem 5.3.1. Let us consider the system $\mathcal{E}$ of all subsets $\mathcal{S}$ and the system $\mathcal{Y}' = \mathcal{E} \setminus \mathcal{X}$. Then the second formula in (5.3) may be rewritten in the appearence $\mathcal{Y} = \{\mathcal{S} \setminus \beta',\ \beta' \in \mathcal{Y}'\}$.
Thus, because of the theorem 5.3.1., we obtain that $\mathcal{X} = \mathcal{Y}^*$. The proposition is proven.

Now we start to construct the systems of stabilizing sets for algorithms VEF with normal weights. As has already been shown, for this class of algorithms stability of prediction $\mathbf{\Pi}_0 : \mathcal{W} = \mathcal{B} \bigsqcup \mathcal{H}$ for an initial decomposition $\mathcal{P}_0 : \mathcal{W} = \mathcal{B}_0 \bigsqcup \mathcal{H}_0$ takes place if $\sigma(\mathcal{B}_0 \bigcup \mathcal{S}) \in \mathcal{L}(\Pi)$ for all $\mathcal{S}' \subseteq \mathcal{S} = \mathcal{B} \subseteq \mathcal{B}_0$. The cone of stability $\mathcal{L}(\Pi)$ in this case is defined by the following system of inequalities:

$$(\sigma,\ \mathcal{X}^\nu) > 0, \quad \nu = 1,\ \cdots,\ M, \tag{5.4}$$

where $\mathcal{X}^\nu$ is the fixed set of N-dimensional vectors defined in chapter ... Thus, the system $\mathcal{X}$ is the maximum set of all $\alpha \subseteq \mathcal{S}$ such that for any $\alpha' \subseteq \mathcal{S} : \alpha \in \mathcal{X}$ we have $\sigma(\mathcal{B}_0 \bigcup \alpha') \in \mathcal{L}(\Pi)$. The last inclusion means that

$$(\sigma(\mathcal{B}_0 \bigcup \alpha'),\ \mathcal{X}^\nu) > 0, \quad \nu = 1,\ \cdots, M\,. \tag{5.5}$$

We fix $\nu \in \{1,\ \cdots, M\}$ and define the system $\mathcal{X}_\nu$ as the maximum set of subsets $\alpha^\nu \subseteq \mathcal{S}$ such that for any $\alpha' \subseteq \mathcal{S},\ \alpha^\nu \in \mathcal{X}_\nu$ we have $(\sigma(\mathcal{B}_0 \bigcup \alpha'),\ \mathcal{X}_\nu) > 0$. Thus, for the system $\mathcal{X}$ of stabilizing sets and for the dual system $\mathcal{Y}$ we obtain:

$$\mathcal{X} = \mathcal{X}_1 \bigotimes \mathcal{X}_2 \bigotimes \cdots \bigotimes \mathcal{X}_m; \quad \mathcal{Y} = \mathcal{X}^* = \mathcal{X}_1^* \bigotimes \cdots \bigotimes \mathcal{X}_m^*\,. \tag{5.6}$$

The algorithm for calculating the system $\mathcal{X}_\nu$ is as follows. We define the inner product $f_\nu(\alpha) = (\sigma(\mathcal{B}_0 \bigcup \alpha),\ \mathcal{X}^\nu)$ and recall that the weights $\sigma_r (r \in \mathcal{V})$ are additive functions over $\mathcal{W}$ (see ...). The function $f_\nu$ is their linear combination and, thus, is also an additive function. We symbolize

$$\mathcal{S}_+^\nu = \{w \in \mathcal{S} : f_\nu(w) > 0\}; \quad \mathcal{S}_-^\nu = \{w \in \mathcal{S} : f_\nu(w) \le 0\}\,. \tag{5.7}$$

Let us consider the numbers $f_\nu(\mathcal{B}_0),\ f_\nu(\mathcal{B}_+^\nu),\ f_\nu(\mathcal{S}_-^\nu)$. If $f_\nu(\mathcal{B}_0) + f_\nu(\mathcal{S}_-^\nu) > 0$, then the inequality $f_\nu(\alpha) > 0$ holds for all $\alpha \subseteq \mathcal{S}$ so that $\mathcal{X}_\nu$ is the set of all subsets of $\mathcal{S}$ (if this condition is met for all $\nu = 1,\ \cdots,\ M$, then prediction $\mathbf{\Pi}_0$ is stable).

If $f_\nu(\mathcal{B}_0) + f_\nu(\mathcal{S}_+^\nu) + f_\nu(\mathcal{S}_-^\nu) = f_\nu(\mathcal{B}) < 0$, then $\mathcal{X}_\nu$ does not contain any subset from $\mathcal{S}$. Therefore, we have that prediction $\mathbf{\Pi}_0$ is not locally stable. Thus, the only other case we have to consider is when

$$f_\nu(\mathcal{B}_0) + f_\nu(\mathcal{S}_-^\nu) < 0 < f_\nu(\mathcal{B}_0) + f_\nu(\mathcal{B}_-^\nu) + f_\nu(\mathcal{S}_+^\nu) .$$

We define $C_1 = (f_\nu(\mathcal{B}_0) + f_\nu(\mathcal{S}^\nu))$, $C_2 = f_\nu(\mathcal{B}_0) + f_\nu(\mathcal{S}_-^\nu) + f_\nu(\mathcal{S}_+^\nu)$. Thus, $C = C_1 + C_2 = f_\nu(\mathcal{S}_+^\nu)$, where $C$, $C_1$, $C_2 > O$.
At this stage, we construct the monotonous threshold system $\mathcal{X}_\nu' = \{\alpha \subseteq \mathcal{S}_+^\nu : f_\nu(\alpha) \geq C_1\}$. The sought after system $\mathcal{X}_\nu$ is equivalent to $\mathcal{X}_\nu'$ in the sense of the equivalence relation, introduced in 5.1. In the same way, we construct the dual system $\mathcal{Y}_\nu' = \{\mathcal{X}_\nu'\}^*$ on $\mathcal{S}_+^\nu$.According to the proposition 5.3.3, we have that $\mathcal{Y}_\nu' = \{\beta \subseteq \mathcal{S}_+^\nu : f_\nu(\beta) > C_2\}$. Doing this for all $\nu = 1, \ldots, M$ we find the system of stabilizing sets $\mathcal{X}$ and the dual system $\mathcal{Y}$.

Now we consider the case of the algorithms VEF-0 and VEF-1. Here, we may assume that the kernel $Z(\mathcal{P}_0)$ is at the center of the coordinates or, in other words, $\sigma_i(\mathcal{B}_0) \geq 0$, $i = 1, \cdots, k$. Thus, for the algorithm VSF-O, the cone $\mathcal{L}(\mathbf{\Pi})$ is defined by the system of inequalities:

$$\sigma_i(\mathcal{B}) > 0, \quad i = 1, \cdots, n \tag{5.8}$$

and, for the algorithm VEF-1, by the system:

$$\sigma_i(\mathcal{B}) > 0, \quad i = k+1, \cdots, n . \tag{5.9}$$

For arbitrary algorithms VSF, for which the weights $\sigma_i$ are not normal, the difficulty is that the functions $\sigma_i$ are not additive on $\mathcal{W}$ and thus, there is no simple algorithm to verify the condition $\sigma(\mathcal{P}) \in \mathcal{L}(\mathbf{\Pi})$. However, as is shown in [70], $sgn\sigma_i = sgn\sigma_i^{\mathcal{H}}$ in the case of algorithms VEF-0 and VEF-1. Therefore, for checking the inequalities 5.8 and 5.9, we may substitute $\sigma_i$ to $\sigma_i^{\mathcal{H}}$ for $\sigma_i$ and apply the technique which has been applied earlier on the case of normal weights.

## 5.4 Local Stability and Cluster Analysis

We expect that a prediction will provide a good division of the whole set of objects into classes $\mathcal{B}$ and $\mathcal{H}$. As A. Rastzvetaev has shown [147], the classes $\mathcal{B}$ and $\mathcal{H}$ occur as compact groups of objects. This fact establishes a link between theory of stability and cluster analysis. This link is important, because the cluster analysis technique allows us to describe "how compact" a group of objects is. Such compactness may be used in evaluating the prediction stability (a "compact set" here does not mean compact in the usual sense of metrical space).

Herewith, we establish the link for one possible definition of cluster [87, 147]. The fact that the objects $w \in \mathcal{B}$ form a cluster is sufficient for the local stability of the prediction $\mathcal{W} = \mathcal{B} \bigsqcup \mathcal{H}$.

We will consider the algorithms VSF($\mathcal{U}$, $\sigma(r, \mathcal{B}_0)$) with normal weights (see 3.3). First, we show that such algorithms consist of two stages. This follows from the fact that the normal weight is an additive function of the second argument (see 3.3). Therefore

$$\hat{\mathcal{F}}(w) = \sum_{r \in \mathcal{U}: \Phi(w) \in r} \sigma(r, \mathcal{P}) = \sum_{r \in \mathcal{U}: \Phi(w) \in r} \sigma(r, \mathcal{B}_0) = \sum_{r \in \mathcal{U}: \Phi(w) \in r} \sum_{w' \in \mathcal{B}_0} \sigma(r, | w' |) . \tag{5.10}$$

Recall that in (5.10) $\mathcal{P}: \mathcal{W} = \mathcal{B}_0 \bigsqcup \mathcal{H}_0$ is an initial decomposition, and $\Phi$ is the coding. We define

$$\mathcal{V}(w, w') = \sum_{r \in (\mathcal{U}: \Phi(w) \in r} \sigma(r, | w' |) . \tag{5.11}$$

Changing the order of summation in (5.10) and taking into account (5.11) we obtain

$$\hat{\mathcal{F}}(w) = \sum_{w' \in \mathcal{B}_0} \mathcal{V}(w, w') . \tag{5.12}$$

The formulas (5.11) and (5.12) divide an algorithm VSF into two stages. In the first stage, we calculate the matrix of "elementary votes" $\| \mathcal{V}(w, w') \|$, $w, w' \in \mathcal{W}$.

In the second stage, the objects $w' \in \mathcal{B}_0$ "vote" for the fact that $w$ $\mathcal{B}$ using formula (5.12).

Thus if $\Delta$ is a chosen voting threshold, then

$$\mathcal{B} = \left\{ w \in \mathcal{W}: \sum_{w' \in \mathcal{B}_0} \mathcal{V}(w, w') \geq \Delta \right\} . \tag{5.13}$$

Assuming that $\mathcal{B}_0 = \mathcal{B}$ in (5.13) and taking into account definition 5.3.2, we obtain the condition of local stability in the following form

$$\exists\Delta: \mathcal{B} = \left\{ w \in \mathcal{W}: \sum_{w' \in \mathcal{B}} \mathcal{V}(w,\ w') \geq \Delta \right\}. \tag{5.14}$$

We next introduce the difference coefficients used in the application of cluster analysis: $d(w,w') = -\mathcal{V}(w,w'),\ w,w' \in \mathcal{W}$. The matrix $\| d(w,w') \|$ $, w,w' \in \mathcal{W}$ is called the matrix of difference coefficients.

**Definition 5.4.1.** A subset $\mathcal{B} \subseteq \mathcal{W}$ is called a cluster of the threshold value $h$, if

$$\mathcal{B} = \left\{ w \in \mathcal{W}: (|\ \mathcal{B}\ | -1)^{-1} \sum_{w' \in \mathcal{B}} d(w,w') \leq h \right\}. \tag{5.15}$$

**Proposition 5.4.1.** Let $\mathbf{\Pi}: \mathcal{B} \bigsqcup \mathcal{H}$ be a prediction, obtained by some algorithm VSF with normal weights. Then if $\mathcal{B}$ is a cluster, the prediction $\mathbf{\Pi}$ is locally stable.

**Proof.** Let $\mathcal{B}$ be a cluster. Then, for some $h$, equation (5.15) holds. We consider $\Delta' = (|\ \mathcal{B}\ | -1)h$.
Thus,

$$\mathcal{B} = \left\{ w \in \mathcal{W}: \sum_{w' in \mathcal{B}} d(w,w') \leq \Delta' \right\}. \tag{5.16}$$

If we take $\Delta = \Delta'$, and come back to the matrix of elementary votes $\| \mathcal{V}(w,w') \|$, we obtain the condition (5.14) of local stability.
The established link between local stability of prediction and cluster analysis gives an opportunity to control the reliability of results in limit recognition problems by an application of cluster analysis techniques.

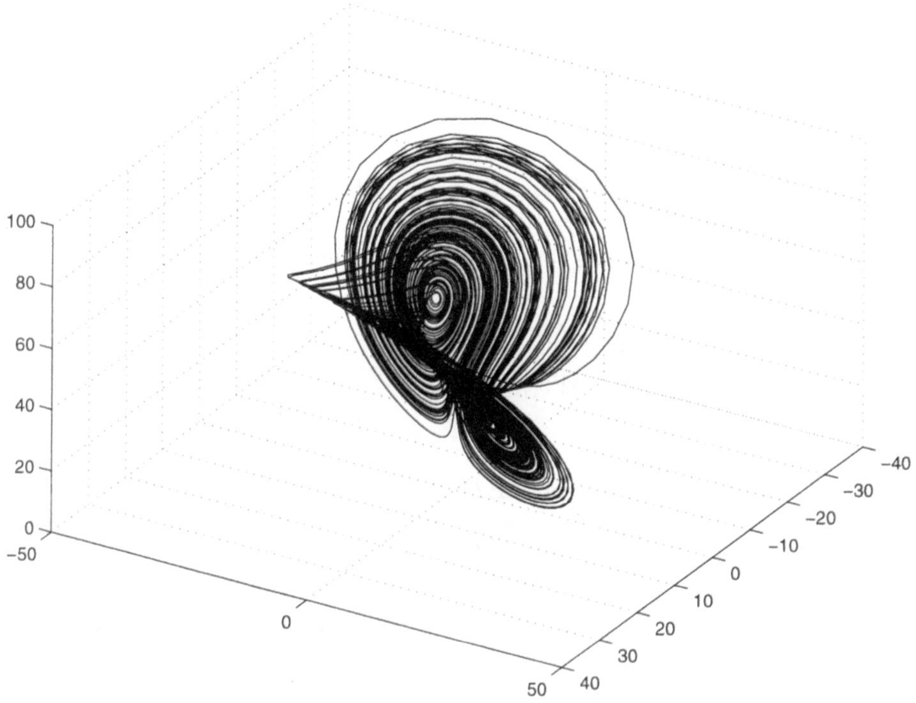


The Lorenz Attractor – The parameters are the same as in page 4, but the perspective representation is different. It shows in the phase space shows 10,000 points discrete phase trajectories (after David AUBERT, 1997).

# Chapter 6

# Investigation of Earthquake-prone Areas as a Limit Pattern Recognition Problem

This chapter begins the process of linking the theory to actual applications. Such applications will be described in detail by the authors in the next book of this series. The application considered here is the determination of strongly earthquake-prone areas and is one of the classical examples of a concrete application of classification pattern-recognition techniques [2, 7, 24, 28, 31, 51, 56, 66, 67, 73, 79, 93, 105, 144, 166].

## 6.1 Formalizing the Recognition of Strongly Earthquake-prone Areas

We consider a seismically active region $\mathcal{R}$ and introduce a threshold of magnitude $M_0$, starting from which $(M \geq M_0)$ earthquakes are considered as strong. The problem is to establish the areas of the region $\mathcal{R}$ where strong earthquake have occured or may occur.

Therefore, by the problem of recognition of strongy earthquake-prone areas we understand a problem of division of a finite set of objects $\mathcal{W}$ situated in the region $\mathcal{R}$ into two classes: a class of objects $w \in \mathcal{W}$ connected with epicenters of earthquakes with $M \geq M_0$ and class of objects connected with only weaker earthquakes ($M < M_0$).

The set $\mathcal{W}$ can be defined in different ways. For example it can be plane squares, circles, morphostructural knots, points on the faults, zones on the borders of continental and oceanic structures, etc. However, for any definition of the objects $\mathcal{W} = \{w\}$ they have to be *a priori* connected with earthquake source zones in the sense of tectonic activity.

The problem of recognition of earthquake-prone areas in such formalization started to be studied in [66], where static pattern-recognition technique was applied.

In this book we go much further and consider the problem as limit (dynamic) pattern recognition problem (see [82]). Such interpretation allows us to start formalizing another problem- how to evaluate the quality of the obtained limit classification and how to make a decision whether it is possible to accept the classification as a prediction or not.

First, we will show that recognition of an earthquake-prone areas is really a limit pattern recognition problem. We have here two classes. Class $\mathcal{B}$ consists of the objects, connected with epicenters of earthquakes with $M \geq M_0$. Class $\mathcal{H}$ contains the rest of the objects $w \in \mathcal{W}$. Through time, objects may transfer from class $\mathcal{H}$ to class $\mathcal{B}$, but not vice-versa. A transfer of $w \in \mathcal{H}$ to class $\mathcal{B}$ takes place at the moment when the first earthquake with $M \geq M_0$ occurs in sufficiently close vicinity to $\mathcal{W}$.

Therefore, decomposition of the objects into two classes depends on time. The graph of permitted transfers has the appearence

$$\mathcal{H} \longrightarrow \mathcal{B}$$

We look for all objects, which at anytime can be connected with epicenters of earthquakes with $M \geq M_0$. This means that we search for the decomposition

$$\mathcal{W} = \lim_{t\to\infty} \mathcal{B}(t) \bigsqcup \mathcal{H}(t)\,, \tag{6.1}$$

where $\mathcal{B}(t)$ and $\mathcal{H}(t)$ are the classes defined in the period up to time $t$. Thus, according to ... this is a limit-pattern recognition problem.

The formalization of prediction can be further generalized in the following natural way. At first we choose several thresholds of magnitudes $M_0 < M_1 < \cdots < M_{r-1}$.

Class $\mathcal{K}_i(t)$, $i = 0, 1, \cdots, r-1$ consists of the objects, for which the maximum observed magnitude (up to time $t$). is $M_{\max}(t) \in [M_{i-1}, M_i]$. The problem is to find the maximum possible magnitudes for all $w \in \mathcal{W}$. In other words, the limit pattern recognition problem is a search of final decomposition

$$\mathcal{W} = \lim_{t \to \infty} \bigsqcup_{i=0}^{r-1} \mathcal{K}_i(t) \tag{6.2}$$

under the condition that the graph of permitted transfers has the appearence

$$\underset{\mathcal{K}_0}{\bullet} \longrightarrow \underset{\mathcal{K}_1}{\bullet} \longrightarrow \bullet \cdots \longrightarrow \underset{\mathcal{K}_{r-2}}{\bullet} \longrightarrow \underset{\mathcal{K}_{r-1}}{\bullet}$$

For prediction of strong earthquakes-prone areas, we have the following given observations:

**1** Known epicenters of earthquakes with $M \geq M_0$ divided into two types:

**1a** Epicenters of earthquakes with $M \geq M_0$, for which there are obvious unique objects connected with those epicenters. The set of such objects we symbolize by $\mathcal{B}_0$ and call the learning class of high seismic objects.

**1b** Epicenters of earthquakes with $M \geq M_0$, for which it is known that an epicenter is connected with a group of objects $w \in \mathcal{W}$, but it is not clear with which particular object from the group it is connected. Such groups of objects we symbolize by $\mathcal{B}_1, \cdots, \mathcal{B}_k$.

**2** The set of parameters $X_1(w), \cdots, X_m(w)$ which describes the objects of recognition. Usually geological, geophysical and geomorphological parameters are used to describe the objects.
The functions $X_1(w), \cdots, X_m(w)$ allows us to represent the objects $w \in \mathcal{W}$ in the appearance of binary vectors *e.g.* to construct the mapping $\phi : \mathcal{W} \to \Omega_n$. An example of such discretization and coding is the algorithm of coding by thirds. It operates in the following way.
We fix a subset $\mathcal{C} \subseteq \mathcal{W}$ and discretize the parameter $X_i$. To do this, the image $X_i$ is represented as three intervals such that the inequalities $X^j < X_i < X^{j+1}$ will distinguish from the set $\mathcal{C}$ roughly the same number of objects. We will give the same value to $X_i$ in the limits of each interval. Thus we can code $X_i$ in the following way

$$\begin{array}{ll} 11, & \text{if } X_i(w) > X^2 \\ 01, & \text{if } X' < X_i(w) \leq X^2 \\ 00, & \text{if } X_i(w) < X^1 \end{array} \tag{6.3}$$

It is clear that the coding (6.3) (defined by the choice of $\mathcal{C}$ applied then and for the objects $w \overline{\in} \mathcal{C}$. As the result of such coding, the values of parameters are separated into three groups: "small", "medium" and "large" (or into any other number of groups). As the result of the coding, we define the mapping $\phi : \mathcal{W} \rightarrow \Omega_n$ which allows us to apply the algorithms described above.
We denote by $\mathcal{H}_0 \subseteq \mathcal{W} \setminus \left[\bigcup_{j=1}^{k} \mathcal{B}_j\right]$ some subset of objects not connected to the present moment of time with epicenters of known strong earthquakes. The set $\mathcal{H}_0$ we can consider as learning material of the class $\mathcal{H}$. Thus, the subset $\mathcal{W}_0 = \mathcal{B}_0 \bigsqcup \mathcal{H}_0$ may be considered an initial decomposition. If the control groups are absent, then $\mathcal{H}_0 = \mathcal{W} \setminus \mathcal{H}_0$, and $\mathcal{W}_0 = \mathcal{W}$.
Therefore, we may apply VSF algorithms to search for the prediction classification. Let

$$\mathcal{O}(t) = (\mathcal{B}_0, \mathcal{H}_0),\ \mathcal{W} = \mathcal{W}_0 \text{ and} \mathcal{A}(0(t), \mathcal{C}(t)) : \mathcal{W} = \mathcal{B} \bigsqcup \mathcal{H} \tag{6.4}$$

be the prediction, obtained by VSF algorithm A.
The stability condition for the prediction (6.4) requires that any sequence of earthquakes which will appear in the future will not change the classification (6.4).
In this case the condition of stability can be reformulated in the following illustrative way. We introduce the ordering $\nu : w_1, \cdots, w_k; k = \mid \mathcal{S} \mid$ on the set $\mathcal{S} = \mathcal{B} \setminus \mathcal{B}_0$. This means that we have introduced a time sequence of the earthquakes ?-markedly the objects where, according to the prediction, strong earthquakes may occur and not known to the present moment of time $t$. In other words, an epicenter potentialy connected with the object $w_j$ appears earlier than an epicenter, connected with the object $w_{j+1}$.
Let $\mathcal{O}(j) = (\mathcal{B}_0(j), \mathcal{H}_0(j))$ be the learning material which corresponds to the epicenters, connected with the objects $w_1, \cdots, w_j$, that have already appeared up to time $t$. In other words

$$\mathcal{B}_0(j) = \mathcal{B}_0 \bigsqcup \{w_1, \cdots, w_j\}, \quad \mathcal{H}_0(j) = \mathcal{H}_0 \setminus \{w_1, \cdots, w_j\}.$$

Thus, for any order $\nu$ and for any $j = 1, \cdots, k$ there are values of free parameters $c(j)$ such that the classification

$$\mathcal{A}(o(j), c(j)) : \mathcal{W} = \mathcal{B}(j) \bigsqcup \mathcal{H}(j) \tag{6.5}$$

is equal to (6.4).
If the condition (6.5) is satisfied only for $j = k$ then the prediction is locally stable.
In some problems for the determination of strong earthquake-prone areas, the condition $\mathcal{W} = \mathcal{W}_0$ does not hold. In this case, we can restrict the stability condition to the subset $\mathcal{W}_0 \subset \mathcal{W}$ and use $\mathcal{U} = \mathcal{W} \setminus \mathcal{W}_0$ as a control set. If such a prediction is stable we also gain an argument for the reliability of the obtained prediction.

## 6.2 Correspondance between Prediction and Classification Problems

As stated in 6.1 by the problem of prediction of earthquake-prone areas, we mean the problem of dividing the set $\mathcal{W}$ into two classes : 1) the $\mathcal{B}$ class of objects $w \in \mathcal{W}$ connected with known or possible epicenters having $M \geq M_0$ and 2) the $\mathcal{H}$ class of objects connected only with earthquakes $M < M_0$.
This is a limit-pattern recognition problem. In particular, this means that we do not have sample objects of the class $\mathcal{H}$. In fact, the places where earthquakes having $M \geq M_0$ are impossible should be identified as one of the results of the problem. Thus, *a priori* such places are unknown.
At the same time, the algorithms we use perform the learning procedure on both classes $\mathcal{B}_0$ and $\mathcal{H}_0$ (but not in a symmetrical way). This requires us to divide the process of solving the problem into two stages:

1. To formulate a classification problem for which we may exactly define the notion of it's solution. (such a solution will be defined using the classification algorithms constructed in chapter ...)

2. evaluation of whether an obtained solution of the classification problem is, in fact, the sought after prediction of the earthquake-prone area.

We define a formal classification problem based on the nature of a given problem of prediction. Notice, first of all, that the final classification may be obtained by using the given coding map $\Phi : \mathcal{W} \to \Omega_n$, as well as by using the reduced vector-representation of objects. In other words, we consider the mapping of projection $p : \Omega_n \to \Omega_m$ $m < n$ and represent the objects $w \in \mathcal{W}$ by linear vectors $\omega = (p \circ \Phi)(\mathcal{W})$ of dimension $m < n$.

We call the procedure of recognition $g^*(\mathcal{O}, p, \mathcal{A})$ the entire process of obtaining the classification (6.1), which consists of reducing the vectors dimension $p$ and applying a classification algorithm $\mathcal{A}$ with learning material $\mathcal{O}$. We write for this :

$$g^*(\mathcal{O}) = g^*(\mathcal{O}, p, \mathcal{A}) : \mathcal{W} = \mathcal{B} \bigsqcup \mathcal{H}. \tag{6.6}$$

We take into account all classifications (6.1) such that:

1. $\mathcal{B}_0 \subset \mathcal{B}$. In other words, any object $w \in \mathcal{B}_0$, connected with a known epicenter of an earthquake with $M \geq M_0$, is classified as a potentialy dangerous object.

2. For every $\mathcal{B}_i$, $i = 1, \cdots, k$, there is an object $w \in \mathcal{B}_i$ such that $w \in \mathcal{B}$. In other words, in each group of objects, connected wuth a given epicenter ($M \geq M_0$) not used in learning process, there is an object from the class $\mathcal{B}$ of "dangerous" objects.

3. $| \mathcal{B} | \leq \beta | \mathcal{W} |$, where $\beta$ is the *a priori* probability that an object belongs to the class $\mathcal{B}$ as $t \to \infty$. In other words, the classification must be sufficiantly non-trivial.

4. The classification (6.1) is the stable solution of the limit pattern recognition problem.

The problem of obtaining a classification (6.1) which satisfies the conditions (2.4) is called the classification problem, (for the problem of predicting strong earthquake-prone areas). A solution of this problem is defined as any classification (6.1) which satisfies the conditions (2.4). If there is no solution, then we substitute the condition 4 for the condition of local stability.
The formulation of the problem of prediction of earthquake-prone areas depends on one free parameter $\mathcal{M}_0$ the threshold of magnitude. The corresponding classification problem has two free parameters:

1. Threshold of magnitude $M_0$, which defines the learning material $\mathcal{O}$ and the control groups $\mathcal{B}_1, \cdots, \mathcal{B}_k$.

2. *A priori* evaluation $\beta$ of the probability of the fact that some object $w$ belongs to the class $\mathcal{B}$ in the decomposition

$$\mathcal{W} = \mathcal{B} \bigsqcup \mathcal{H} = \lim_{t \to \infty} \{\mathcal{B}(t) \bigsqcup \mathcal{H}(t)\}. \tag{6.7}$$

The choice of $M_0$ and $\beta$ to formulate the classification problem may be done by expert estimations taking into account optimizations of reliability functions (see [90, 160]).
As follows from the formulation of the classification problem, it's solutions may be considered as possible predictions of strong earthquake-prone areas. More precisely, from the conditions 2.4 it follows that such solutions provide non-trivial, stable predictions without obvious errors (on the learning material).
The evaluation of wether to accept a solution of the classification problem as the desired prediction is both important and complicated. To tackle this problem based on the limited information available, we need to formulate additional conditions and choose only solutions that satisfy those conditions in the best way. Such a way of tackling the problem looks natural since we cannot formulate any sufficient condition for the prediction acceptance.
The subspace $(p \circ \Phi)(\mathcal{W})$ which corresponds to this solution represents the subspace of the most informative parameters

$$X_{i_1}(w), \quad \cdots, X_{i_k}(w)$$

in terms of which the algorithms form criteria of high ($M \geq M_0$) seismicity.
The proposed way of doing prediction of strong earthquake-prone areas may be described by the following scheme:

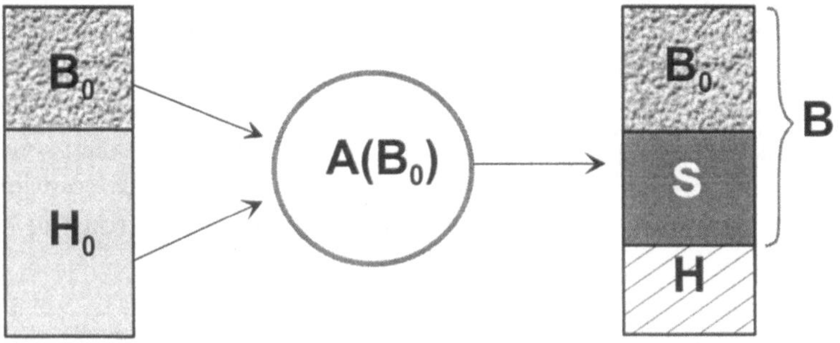


*a)*

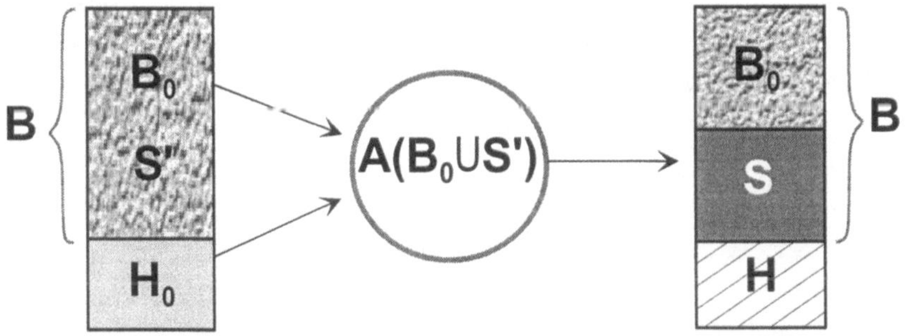


*b)*

Stability condition for dynamic pattern recognition problem - $\mathcal{A}(\mathcal{B}_0)$- a VSF classification algorithm with learning material a) $\mathcal{W} = \mathcal{B}_0 \bigsqcup \mathcal{H}_0$- initial decomposition (learning material) b) $\mathcal{S}'$ any subset $\mathcal{S}' \subseteq \mathcal{S}$, $\mathcal{W} = [\mathcal{B}_0 \bigcup \mathcal{S}] \bigsqcup \mathcal{H} = \mathcal{B} \bigsqcup \mathcal{H}$ - final classification (prediction).

# Chapter 7

# Control Experiments for Evaluating Classification Reliability

## 7.1 Comparison of Quality Solutions for Real and Random Learning Materials

To compare the quality of classifications, it is natural to apply a Monte-Carlo approach which in our case means to apply the recognition procedure to random learning material generated in a way that is analogous to the real one [82]. To realize this approach, we need to formalize the recognition procedure. In other words we need to construct an algorithm $g(\mathcal{O}, \mathcal{A})$ which models the recognition procedure $g^{\star}(\mathcal{O}, p, \mathcal{A})$ described above and which, automatically, gives us a certain classification for a given learning material $\mathcal{O}$ and a given algorithm $\mathcal{A}$. We call $g(\mathcal{O}, \mathcal{A})$ a modeling algorithm. Herein, we describe the constructions of modeling algorithms for the following four cases: $1)\mathcal{A} = \mathcal{J} = VEF - 0$, $2)\mathcal{A} = \mathcal{J}_1^{\sigma} = VEF - 1$, $3)\mathcal{A} = \mathcal{B}$ – "Bayes" algorithm, and, $4)\mathcal{A} = \mathcal{K}$ – "CORA-algorithm" (see Chapter 1).

To begin, we model the choice of the most informative parameters. For the real procedure $g^{\star}(\mathcal{O}, p, \mathcal{A})$, this is done by analysis of one-dimensional distributions of parameters and preliminary results of recognition.

In the modeling algorithm, we divide the components of vectors into $n$-groups (such groups correspond to natural divisions of the parameters; for example, a group of parameters of altitude, gravity field descriptions, etc.) From the group $\mathcal{I}_l$ of the component's numbers, we choose the number $l^{\star}$ such that

$$| \mathcal{P}_{l^*}(\mathcal{B}_0) - \mathcal{P}_{l^*}(\mathcal{H}_0) |= \max_{i \in \mathcal{I}_l} | \mathcal{P}_i(\mathcal{B}_0) - \mathcal{P}_i(\mathcal{H}_0) |, \qquad (7.1)$$

where $\mathcal{P}_i(\mathcal{B}_0)$ and $\mathcal{P}_i(\mathcal{H}_0)$ are the frequencies of occurence of 1 as defined in 3.2. Then, we construct new vectors $\boldsymbol{\omega} = (\boldsymbol{\omega}(\mathbf{1}), \boldsymbol{\omega}(\mathbf{2}), \ldots, \boldsymbol{\omega}(\boldsymbol{m}))$ which consist only of the components with numbers $l^* \in \mathcal{I}_l$ for, $l = 1, \ldots, m$. In other words, for each group we take the parameter which divides the classes $\mathcal{B}_0$ and $\mathcal{H}_0$ in the best way in the sense of occurence of 1 in binary components of parameters.
The next step is application of one of the algorithms $\mathcal{A} = \mathcal{J}, \mathcal{J}_1^\sigma, \mathcal{B}$, or $\mathcal{K}$. To do this, we need to formalize the choice of free parameters of the algorithms. This is done as follows.
Considering classifications $g(\mathcal{O}, \mathcal{A}) : \mathcal{W} = \mathcal{B}(g) \bigsqcup \mathcal{H}(g)$, we require that $| \mathcal{B}(g) | \leq \beta | \mathcal{W} |$ where $\beta$ is the *a priori* evaluation of the probability of an object belonging to the class $\mathcal{B}(t \to \infty)$, (see 6.2). This requirement leads to the following choice of $\Delta \in \mathcal{C}$.

$$\Delta = \max \{r : | \{w \in \mathcal{W} : f(w) \leq r\} | \leq \beta | \mathcal{W} |\} \qquad (7.2)$$

where

$$f(w) = f_1(w) = \begin{cases} \rho(w, z) & \text{if } \mathcal{A} = \mathcal{J} \\ \rho^\sigma(w, z) & \text{if } \mathcal{A} = \mathcal{J}_1^\sigma \\ 1 - \mathcal{P}\{w \in \mathcal{B} \mid w = \{X_1, \cdots, x_n\}\}, & \text{if } \mathcal{A} = \mathcal{B} \\ -\Delta(w) & \text{if } \mathcal{A} = \mathcal{K}. \end{cases} \qquad (7.3)$$

In other words, to minimize the number of errors on the sets $\mathcal{B}_i$, $i = 0, 1, \cdots, k$, we include into the class $\mathcal{B}$ the maximum number of objects permitted by the conditions of the classification problem. This becomes more obvious if for $\mathcal{A} = \mathcal{B}$ or $\mathcal{K}$ we rewrite 7.2 in the form

$$\Delta = \min \{r : | \{w \in \mathcal{W} : \varphi(w) \geq r\} | \leq \beta | \mathcal{W} |\} \qquad (7.4)$$

where

$$\varphi(w) = \varphi_1(\omega) = \begin{cases} \mathcal{P}\{\omega \in \mathcal{B} \mid \omega = (x_1, \cdots, x_n)\}, & \mathcal{A} = \mathcal{B} \\ \Delta(\omega), & \mathcal{A} = \mathcal{K} \end{cases} \qquad (7.5)$$

As is shown in 3.7 for the algorithm $\mathcal{B}$ a change of $p_{\mathcal{B}}$ does not change the order of objects according to the *a posteriori* probability $\mathcal{P}\{\omega \in \mathcal{B} \mid \omega = (X_1, \cdots, X_n)\}$.

Thus the free parameter $p_{\mathcal{B}}$ in $g(\mathcal{O},\ \mathcal{F})$ is defined as an arbitrary but fixed number. Thus, we have completely defined the natural choice of free parameters in the modeling algorithms $g(\mathcal{O},\ \mathcal{A})$, $\mathcal{A}=\mathcal{J},\ \mathcal{J}_1^{\sigma},\ \mathcal{B}$.
According to 3.4 in a real recognition procedure $g^{\star}(\mathcal{O},\ p,\ \mathcal{K})$, the free parameters $K_{\mathcal{B}},\ \widetilde{K_{\mathcal{B}}},\ K_{\mathcal{H}},\ \widetilde{K_{\mathcal{H}}}$ are choosen by direct analysis of a learning material $\mathcal{O}$ and by characteristic features obtained from the procedure. In the algorithm $g(\mathcal{O},\ \mathcal{K})$, we model this by introducing the following functions

$$\begin{array}{ll} K_{\mathcal{B}} = [\alpha_{\mathcal{B}}\cdot \mid \mathcal{B}_0 \mid] & \widetilde{K_{\mathcal{B}}} = [\tilde{\alpha}_{\mathcal{B}}\cdot \mid \mathcal{H}_0 \mid] \\ K_{\mathcal{H}} = [\alpha_{\mathcal{H}}\cdot \mid \mathcal{H}_0 \mid] & \widetilde{K_{\mathcal{H}}} = [\tilde{\alpha}_{\mathcal{H}}\cdot \mid \mathcal{B}_0 \mid] \end{array} \tag{7.6}$$

The constants $\alpha_{\mathcal{B}},\ \tilde{\alpha}_{\mathcal{B}},\ \alpha_{\mathcal{H}},\ \tilde{\alpha}_{\mathcal{H}}$ depend on the specific problem. To simplify the model we often take $\alpha=\alpha_{\mathcal{B}}=\alpha_{\mathcal{H}},\quad \tilde{\alpha}=\tilde{\alpha}_{\mathcal{B}}=\tilde{\alpha}_{\mathcal{H}}$.
The constants $\alpha$ and $\tilde{\alpha}$ (contrary to $K_{\mathcal{B}},\ \widetilde{K_{\mathcal{B}}},\ K_{\mathcal{H}}$ and $\widetilde{K_{\mathcal{H}}}$) do not depend on the learning material $\mathcal{O}$ and are choosen in a way such the classifications $g^{\star}(\mathcal{O},\ p,\ \mathcal{K})$ and $g(\mathcal{O},\ \mathcal{K})$ are as close as possible.
Applying a modeling algorithm $g(\mathcal{O},\ \mathcal{A})$ we obtain a decomposition

$$g(\mathcal{O},\ \mathcal{A})\colon \mathcal{W} = \mathcal{B}(g) \bigsqcup \mathcal{H}(g) \tag{7.7}$$

which may not be a solution of the classification problem.
But in any case because of (7.2) the classification (7.7) is sufficiently close to a solution.
Now we come to the comparison of the real and random problems. We use here our usual symbolisation where $\mathcal{O}$ is the learning material and $g(\mathcal{O})$ is the modeling algorithm which models the recognition procedure $g^{\star}(\mathcal{O},p,\mathcal{A})$. Let $\mathcal{U}$ be a set of objects $w \overline{\in} \mathcal{W}_0$ with known classification $\mathcal{U}=\lim_{t\to\infty}\mathcal{Y}(t)\bigsqcup\mathcal{Z}(t)=\mathcal{Y}\bigsqcup\mathcal{Z}$. We call $\mathcal{U}$ a control set.
In earthquake-prone area problems we may have in principle only control sets of the appearance $\mathcal{U}=\mathcal{Y}\bigsqcup\emptyset$, where $w\in\mathcal{Y}$ are the objects, connected with epicenters of known earthquakes with $M\geq M_0$.
It is natural to consider two types of fee for the errors on the control set $\mathcal{U}$ while applying the algorithm $g(\mathcal{O})\colon\mathcal{U}=\mathcal{B}_{\mathcal{U}}\bigsqcup\mathcal{H}_{\mathcal{U}}$.

1. Number of the classification errors

$$r(g(\mathcal{O}),\ \mathcal{U}) = \mu(g(\mathcal{O}),\ \mathcal{U}) = \mid (\mathcal{Y} \bigcap \mathcal{H}_{\mathcal{U}} \mid \tag{7.8}$$

2. Frequency of classification errors

$$r(g(\mathcal{O}),\ \mathcal{U}) = \nu(g(\mathcal{O}),\ \mathcal{U}) = \frac{\mid (\mathcal{Y} \cap \mathcal{H}_{\mathcal{U}}) \mid}{\mid \mathcal{Y} \mid} \tag{7.9}$$

The suggested procedure gives an opportunity to compare the values $r(g(\mathcal{O}),\ \mathcal{B}_0)$ and $\mathcal{M}_r(g(\tilde{\mathcal{O}}),\ \tilde{\mathcal{B}}_0)$ where $\tilde{\mathcal{O}} = (\tilde{\mathcal{B}}_0,\ \widetilde{\mathcal{H}}_0)$ is the random learning material generated in analogy to $\mathcal{O}$, and $\mathcal{M}_r(g(\tilde{\mathcal{O}}), \tilde{\mathcal{B}}_0)$ is the mathematical expectation of the random function $r(g(\tilde{\mathcal{O}}),\ \tilde{\mathcal{B}}_0)$. We consider the experiment as successful if $r(g(\mathcal{O}),\ \mathcal{B}_0)$ is significantly smaller than $\mathcal{M}_r(g(\tilde{\mathcal{O}}))$ and the probability of the inequation

$$|\, r(g(\tilde{\mathcal{O}}),\ \tilde{\mathcal{B}}_0) - \mathcal{M}_r(g(\tilde{\mathcal{O}}),\ \tilde{\mathcal{B}}_0)\, | \ \geq\ |\, r(g(\mathcal{O}),\ \mathcal{B}_0) - \mathcal{M}_r(g(\tilde{\mathcal{O}}),\ \tilde{\mathcal{B}}_0)\, | \tag{7.10}$$

is sufficiently small.

If the experiment is successful then we may consider that the classification (7.7) represents a real law of division of the objects $w \in \mathcal{W}$ corresponding to the learning classes $\mathcal{B}_0$ and $\mathcal{H}_0$. If we also have a solution of the classification problem $g^\star(\mathcal{O})\colon \mathcal{W} = \mathcal{B} \bigsqcup \mathcal{H}$, then

$$r(g^\star(\mathcal{O}), \mathcal{B}_0) = 0 \text{ and } r(g^\star(\mathcal{O}), \mathcal{B}_0) \leq r(g(\mathcal{O}), \mathcal{B}_0).$$

Thus, a positive result of the experiment of comparison between the real problem and the random one is, in some sense, a necessary condition of acceptance of the classification problem solution as the desired prediction.

The scheme of the whole experiment is given on the figure.

This scheme may be realized as follows :

1. We consider the random number $\xi$, which takes the values "$\mathcal{B}$", "$\mathcal{H}$", "$\mathcal{E}$", with the following probabilities :

$$\xi = \begin{cases} \mathcal{B}, & p = |\, \mathcal{B}_0 \,| \,/\, |\, \mathcal{W} \,| \\ \mathcal{H}, & p = |\, \mathcal{H}_0 \,| \,/\, |\, \mathcal{W} \,| \\ \mathcal{E}, & p = |\, \mathcal{W} \setminus (\mathcal{B}_0 \bigcup \mathcal{H}_0) \,| \,/\, |\, \mathcal{W} \,| \end{cases} \tag{7.11}$$

We assign integer numbers to the elements of the set $\mathcal{W} = \{w_1,\ \cdots,\ w_{N-1},\ w_N\}$ $N = |\, \mathcal{W} \,|$. Then, $N$ independent realizations of $\xi$, which may be represented as the vectors $\hat{\xi} = (\xi_1,\ \cdots,\ \xi_N)$, define the random learning material

$$\begin{aligned} \tilde{\mathcal{B}}_0 &= \tilde{\mathcal{B}}_0(\hat{\xi}) &&= \{w_k \in \mathcal{W}\colon \xi_k = \mathcal{B}\} \\ \widetilde{\mathcal{H}}_0 &= \widetilde{\mathcal{H}}_0(\hat{\xi}) &&= \{w_k \in \mathcal{W}\colon \xi_k = \mathcal{H}\} \\ \tilde{\mathcal{O}} &= (\tilde{\mathcal{B}}_0,\ \widetilde{\mathcal{H}}_0). \end{aligned} \tag{7.12}$$

2. Apply the modeling algorithm. Let $g^{\star}(\mathcal{O})$ be the given procedure of recognition. Upon applying the modeling algiritthm $g(\tilde{\mathcal{O}})$ we obtain the classification

$$g(\tilde{\mathcal{O}}): \mathcal{W} = \tilde{\mathcal{B}}(\hat{\xi}) \bigsqcup \tilde{\mathcal{H}}(\hat{\xi}) . \tag{7.13}$$

On the other hand, $g(\mathcal{O})$ applied to the real learning material gives the classification

$$g(\mathcal{O}): \mathcal{W} = \mathcal{B}(g) \bigsqcup \mathcal{H}(g) \tag{7.14}$$

3. Calculation of the fees (7.8) and (7.9) for the classifications (7.13) and (7.14). For (7.13) we symbolize the fees (7.8) and (7.9) correspondingly by $\mu(\hat{\xi})$ and $\nu(\hat{\xi})$.

4. Calculations of mathematical expectations and dispersions of the fees:

   By a general set we will understand the set of all classifications (7.13) or the whole set of vectors $\hat{\xi}$.

   Then the fee $r(\hat{\xi}) = r(g(\tilde{\mathcal{O}}), \tilde{\mathcal{B}}_0)$ is a quantitative feature of this general set. Generating $n$-times the random learning material we obtain $n$ samples of this quantitative feature : $r(\hat{\xi}_1), \cdots, r(\hat{\xi}_n)$.

   We estimate the mathematical expectation by the usual formula

$$\bar{r}(\hat{\xi})_{\mathcal{B}} = \frac{1}{n} \sum_{i=1}^{n} r(\hat{\xi}_i) . \tag{7.15}$$

   At the next stage we estimate the dispersion by the formula

$$\mathcal{D}r(\hat{\xi})_{\mathcal{B}} = \frac{1}{n} \sum_{i=1}^{n} (r(\hat{\xi} - \bar{r}(\hat{\xi})_{\mathcal{B}})^2 . \tag{7.16}$$

   Strictly speaking, to estimate the dispersion we need to calculate the so-called corrected dispersion

$$s^2 r(\hat{\xi})_{\mathcal{B}} = \frac{1}{n-1} \sum_{i=1}^{n} (r(\hat{\xi} - \bar{r}(\hat{\xi})_{\mathcal{B}})^2 . \tag{7.17}$$

   However, according to (6.17), if $n > 30$ then (7.17) does not differ significantly from the simplier estimation

$$\mathcal{D}r(\hat{\xi})_\Gamma = \frac{1}{n}\sum_{\xi}\left(r(\hat{\xi}) - \mathcal{M}r(\hat{\xi})\right)^2. \qquad (7.18)$$

We generate $\hat{\xi}$ not less than 50 times and use the formula

$$\begin{aligned} \sigma r(\hat{\xi})_\mathcal{B} &= \sqrt{\frac{1}{n}\sum_{i=1}^{n} r^2(\hat{\xi}_i) - (\frac{1}{n}\sum_{i=1}^{n} r(\hat{\xi}_i))^2} \\ &= \sqrt{\overline{r^2(\hat{\xi})_\mathcal{B}} - [\overline{r}(\hat{\xi})_\mathcal{B}]^2}. \end{aligned} \qquad (7.19)$$

5. Compare the fees for the real and random learning materials

   Finally, we compare the values of $\mu$ and $\overline{\mu}(\hat{\xi})_\mathcal{B}, \nu$ and $\overline{\nu}(\hat{\xi})_\mathcal{B}$. To do this we use the estimations of probability $\mathcal{Q}_\mu$ and $\mathcal{Q}_\nu$ to obtain the difference $\mid r - \overline{r}(\hat{\xi}) \mid$ in some random realization. Here we assume that the fees $r(\hat{\xi})$ are normally distributed. This gives us an opportunity to calculate the estimations

$$\mathcal{P}\left\{\mid r(\hat{\xi}) - \overline{r}(\hat{\xi})_\mathcal{B} \mid \geq \mid r - \overline{r}(\hat{\xi})_\mathcal{B} \mid\right\} \leq \mathcal{Q}r;\ r = \mu, \nu$$

   using, for example, the tables []. We consider the result of this experiment as successful if $\mathcal{Q}_\mu$ and $\mathcal{Q}_\nu$ are sufficiently small.

## 7.2 Estimating the Probability of the Classification Error

A small value of the classification error

$$g^\star(\mathcal{O}, p, \mathcal{A}) : \mathcal{W} = \mathcal{B}\bigsqcup\mathcal{H} \qquad (7.20)$$

is an obvious necessary condition for a classification problem solution (7.2.1) to be a prediction. At the same time, estimation of the classification error is a difficult mathematical problem which has at least several different approaches. Here we introduce a method of evaluation of the classification error which is based on some assumptions connecting classifications obtained for real and random learning materials [82, 90].

We define a random control set in the same way as the random learning material $\tilde{\mathcal{O}}$ (see 7.1). We consider $\mathcal{U} = \mathcal{Y} \bigsqcup \mathcal{Z}$ as a pair $(\mathcal{Y},\ \mathcal{Z})$, $\mathcal{Y} \subset \mathcal{U}$, $\mathcal{Z} \subset \mathcal{U}$ and construct $(\tilde{\mathcal{Y}},\ \tilde{\mathcal{Z}})$ by generating $\mid \mathcal{U} \mid$ times the random value

$$\xi = \begin{cases} 1 & , \quad p_1 = \dfrac{\mid \mathcal{Y} \mid}{\mid \mathcal{U} \mid} \\ 0 & , \quad p_2 = \dfrac{\mid \mathcal{Z} \mid}{\mid \mathcal{U} \mid} \end{cases} \tag{7.21}$$

An object $u \in \mathcal{U}$ belongs to $\tilde{Y}$, if $\xi = 1$ and $u \in \tilde{\mathcal{Z}}$, if $\xi = 0$ in the corresponding generation. We symbolize $\tilde{\mathcal{U}} \bigsqcup \tilde{\mathcal{Z}}$. Thus, we obtain random control sets which are generated in analogy to the real control set.
Following [145] we introduce the assumption

$$\nu(g(\mathcal{O}),\ \mathcal{U}) - \nu(g(\mathcal{O}),\ \mathcal{B}_0) \leq M\nu(g(\tilde{\mathcal{O}}),\ \tilde{\mathcal{U}}) - M\nu(g(\tilde{\mathcal{O}}),\ \tilde{\mathcal{B}}_0) \tag{7.22}$$

Here $\mathcal{U}$ is the control set and

$$\begin{aligned} \nu(g(\mathcal{O}),\ \mathcal{U}) &= \frac{\mid [\mathcal{Y} \cap \mathcal{H}_{\mathcal{U}}(g(\mathcal{O}))] \bigcup [\mathcal{Z} \cap \mathcal{B}_{\mathcal{U}}(g(\mathcal{O}))] \mid}{\mid \mathcal{U} \mid} \\ \nu(g(\tilde{\mathcal{O}}),\ \tilde{\mathcal{U}}) &= \frac{\mid [\tilde{\mathcal{Y}} \cap \mathcal{H}_{\mathcal{U}}(g(\tilde{\mathcal{O}}))] \bigcup [\tilde{\mathcal{Z}} \cap \mathcal{B}_{\mathcal{U}}(g(\mathcal{O}))] \mid}{\mid \tilde{\mathcal{U}} \mid} \end{aligned} \tag{7.23}$$

In the same way

$$\nu(g^{\star}(\mathcal{O}),\ \mathcal{U}) = \frac{\mid (\mathcal{Y} \cap \mathcal{H}_{\mathcal{U}}(g^{\star})) \bigcup (\mathcal{Z} \cap \mathcal{B}_{\mathcal{U}}(g^{\star}))}{\mid \mathcal{U} \mid}. \tag{7.24}$$

While calculating the mathematical expectation in (7.22) we consider all possible generations $(\tilde{\mathcal{O}},\ \tilde{\mathcal{U}})$.
The assumption (7.22) means that when we shift from the control on "familiar" material $(\mathcal{B}_0,\ \tilde{\mathcal{B}}_0)$ to the control on "non familiar" material $(\mathcal{U},\ \tilde{\mathcal{U}})$ we get a smaller change of the fee for the real material on average than for the random one. The assumption (7.22) becomes even more clear if $\mathcal{Q}_\nu$ is sufficiently small. In this case $g(\mathcal{O})$ gives a classification with a fee, which is possible to obtain, on average, for the random learning material only with a very low probability. Thus, $g(\mathcal{O})$ divides the objects $w \in \mathcal{W}$ according to a law which possibly divides as well the objects $u \in \mathcal{U}$. Therefore, we may expect that the difference in the left part of (7.22) is sufficiently small to be considered zero.

We also introduce the following assumption: starting from the control on "familiar" material (learning material $\mathcal{B}_0$) to the control on "not familiar" material (control set $\mathcal{U}$) a modeling algorithm loses not less than the actual recognition procedure:

$$\nu(g^\star(\mathcal{O}),\ \mathcal{U}) - \nu(g^\star(\mathcal{O}),\ \mathcal{B}_0) \leq \nu(g(\mathcal{O}),\ \mathcal{U}) - \nu(g(\mathcal{O}),\ \mathcal{B}_0)\,. \tag{7.25}$$

The assumption (7.25) is also rather natural. By definition, a modelling algorithm reflects the disired division of the objects no better than the recognition procedure which it is modeling.
From (7.22) and (7.25) it follows, that

$$\nu(g^\star(\mathcal{O}),\ \mathcal{U}) - \nu(g^\star(\mathcal{O}),\ \mathcal{B}_0) \leq M\nu(g(\tilde{\mathcal{O}}),\ \tilde{\mathcal{U}}) - M\nu(g(\tilde{\mathcal{O}}),\ \tilde{\mathcal{B}}_0). \tag{7.26}$$

and

$$\nu(g^\star(\mathcal{O}),\ \mathcal{U}) \leq \nu(g^\star(\mathcal{O}),\ \mathcal{B}_0) + M\nu(g(\tilde{\mathcal{O}}),\ \tilde{\mathcal{U}}) - M\nu(g(\tilde{\mathcal{O}}),\ \tilde{\mathcal{B}}_0)$$

We define the probability of classification error $\mathcal{P}_{g^\star}$ by the formula

$$\mathcal{P}_{g^\star} = \lim_{|\mathcal{U}|\to\infty} \nu(g^\star(\mathcal{O}),\ \mathcal{U})\,. \tag{7.27}$$

Because (7.27) is the limit of the normal numerical sequence, we obtain from (7.27) (see for example (...)) that

$$\mathcal{P}_{g^\star} \leq \nu(g^\star(\mathcal{O}),\ \mathcal{B}_0) + M\nu(g(\tilde{\mathcal{O}},\ \tilde{\mathcal{U}}) - M\nu(g(\tilde{\mathcal{O}}),\ \tilde{\mathcal{B}}_0)\,. \tag{7.28}$$

Everywhere below we shall consider the case where $\mathcal{U} = \mathcal{U}' = \mathcal{Y} \bigsqcup \emptyset$ (see 7.20). Thus, to accept the solution of the classification problem as a prediction, the probability

$$\mathcal{P}_{g^\star}(\mathcal{B}) = \lim_{|\mathcal{Y}|\to\infty} \nu(g^\star(\mathcal{O}),\ \mathcal{Y})$$

should be sufficiently small.
If in the same way we restrict (7.28) to $\mathcal{U} = \mathcal{U}'$, we obtain

$$\mathcal{P}_{g^\star}(\mathcal{B}) \leq \nu(g^\star(\mathcal{O}),\ \mathcal{B}_0) + M\nu(g(\tilde{\mathcal{O}},\ \mathcal{Y}) - M\nu(g(\tilde{\mathcal{O}}),\ \tilde{\mathcal{B}}_0)\,. \tag{7.29}$$

Now we introduce the third assumption which follows from the randomness of the algorithm $g(\tilde{\mathcal{O}})$ : for a random learning material, a part of the objects classified to one of the classes remains about the same as it is shifted from the learning material to a control set. Formally, this means:

$$M_{\tilde{\mathcal{O}},\mathcal{Y}}\nu(g(\tilde{\mathcal{O}}),\ \mathcal{Y}) = M_{\tilde{\mathcal{O}},\mathcal{Y}}\frac{|\ \mathcal{H}_{\mathcal{U}'}(g(\tilde{\mathcal{O}}))\ |}{|\ \mathcal{Y}\ |} = M_{\tilde{\mathcal{O}}}\frac{|\ \mathcal{H}(g(\tilde{\mathcal{O}}))\ |}{|\ \mathcal{W}\ |}\,. \tag{7.30}$$

Taking into account (7.30) we may rewrite (7.29) in the appearance

$$\mathcal{P}_{g^\star}(\mathcal{B}) \le \nu(g^\star(\mathcal{O}),\ \mathcal{B}_0)+\mid \mathcal{W}\mid^{-1} M \mid \mathcal{H}(g(\tilde{\mathcal{O}})\mid - M\nu(g(\tilde{\mathcal{O}}),\ \tilde{\mathcal{B}}_0)\,. \tag{7.31}$$

Note that if $g^\star$ is a solution of the classification problem, then $\nu(g^\star(\mathcal{O}),\ \mathcal{B}_0) = 0$.

The probability $\mathcal{P}_{g^\star}(\mathcal{B})$ is sufficiently small if the right part of (7.31) is sufficiently small. To calculate the right part of (7.31), we execute many times the recognition for random learning material $\tilde{\mathcal{O}}$. In particular for the values of $l$ large enough we obtain the following estimation:

$$\mathcal{P}_{g^\star}(\mathcal{O},l) = \nu(g^\star(\mathcal{O}),\ \mathcal{B}_0) = \frac{1}{l\mid\mathcal{W}\mid}\sum_{i=1}^{l}\mid \mathcal{H}(g(\tilde{\mathcal{O}}_i))\mid - \frac{1}{l}\sum_{i=1}^{l}\nu(g(\tilde{\mathcal{O}}_i),\ \tilde{\mathcal{B}}_0(\hat{\xi}_i)) \tag{7.32}$$

where $\tilde{\mathcal{O}}_i = (\mathcal{B}_0(\hat{\xi}_i),\ \mathcal{H}_0(\hat{\xi}_i))$ is $i-$th realization of $\tilde{\mathcal{O}}$.

The formula (7.32) gives a practical estimation of the classification error. Thus, the function $\mathcal{P}_{g^\star}(\mathcal{O},l)$ will be called by the function of reliability. A sufficiently small value of (7.32) may be considered as a necessary condition of acceptance of a solution of the classification problem as the desired prediction.

## 7.3 Estimation of Parameters of the Classification Problem using Estimation of Non-randomness and Reliability Functions

In our approach, we construct the prediction from the solutions of the corresponding classification problem. According to 6.2 the definition of a solution

of the classification problem depends on two free parameters: $M_0$ and $\beta$. Initially, these parameters are selected using the observed seismicity in the considered regions. The upper estimation $M_0 \leq M_0^2$ is defined by the applied technique, because for application of the algorithms introduced in chapter 1 we need a sufficiently large number of objects $w \in \mathcal{B}_0$.
On the other hand, we have a natural estimation of $M_0$ values from below: $M_0 \geq M_0^1$. Indeed, with a sufficiently low threshold $M_0$, the epicenters of earthquakes with $M \geq M_0$ may correspond to all the objects $w \in \mathcal{W}$ which makes the problem trivial.
In an analogous way, we construct the initial range $\beta_1 \leq \beta \leq \beta_2$. Thus, we may assume that

$$M_0 \in [M_0^1,\ M_0^2] \quad \text{and}\ \beta \in [\beta_1,\ \beta_2]. \tag{7.33}$$

Here we show how to choose $M_0$ and $\beta$ from the segments (7.33) based on optimizing the non-randomness and reliability functions.
For all $M_0 \in [M_0^1,\ M_0^2]$ let the set $\mathcal{W}$ be eligible for the prediction of strong earthquake-prone areas. Then, each $M_0 \in [M_0^1,\ M_0^2]$ defines both the set of objects $\mathcal{B}_0(M_0)$ connected with epicenters of earthquakes having $M \geq M_0$ as well as the set $\mathcal{H}_0(M_0) = \mathcal{W} \setminus \mathcal{B}_0(M_0)$. Thus we obtain the learning material $\mathcal{O}(M_0) = (\mathcal{B}_0(M_0),\ \mathcal{H}_0(M_0))$.
In the same way as in 6.1, we may construct for $\mathcal{O}(M_0)$ a random value $\xi$, which in this case is a function of $M_0$, $\xi = \xi(M_0)$. It is obvious, that the corresponding learning material $\widetilde{\mathcal{O}}(M_0) = (\widetilde{\mathcal{B}}_0(M_0),\ \widetilde{\mathcal{H}}_0(M_0))$ satisfies the condition $\widetilde{\mathcal{B}}_0(M_0) \bigsqcup \widetilde{\mathcal{H}}_0(M_0) = \mathcal{W}$.
For each $M_0 \in [M_0^1,\ M_0^2]$, we apply the recognition procedure $g^\star(\mathcal{O}_1(M_0), p, \mathcal{A})$ where $\mathcal{O}_1(M_0)$ is learning material close to $\mathcal{O}(M_0)$.
In the formulation of the classification problem, we substitute $g^\star$ by the modeling algorithm $g(\mathcal{O}(M_0), \mathcal{A})$.
As is shown in 7.1, the modeling algorithm $g(\mathcal{O}(M_0), \mathcal{A})$ depends on the choice of $\beta \in (\beta_1,\ \beta_2)$.
Therefore, for each $\mathcal{A}$ and $M_0 \in [M_0^1,\ M_0^2]$ we have a set of modeling algorithms

$$g(\mathcal{O}(M_0), \mathcal{A}) = g(M_0,\ \beta,\ \mathcal{A}) : \mathcal{W} = \mathcal{B}(g(M_0,\ \beta,\ \mathcal{A}))\bigsqcup \mathcal{H}(g(M_0,\ \beta,\ \mathcal{A})). \tag{7.34}$$

At the same time, generation of the random material $\widetilde{\mathcal{O}}(M_0)$ gives us the classification

$$\tilde{g}(M_0, \beta, \mathcal{A}) :\ \mathcal{W} = \mathcal{B}(\tilde{g}(M_0, \beta, \mathcal{A}))\bigsqcup \mathcal{H}(\tilde{g}(M_0, \beta, \mathcal{A})) \tag{7.35}$$

obtained by algorithm $\tilde{g}(M_0, \beta, \mathcal{A}) = g(\tilde{\mathcal{O}}, M_0, \mathcal{A})$.

Therefore, for any algorithm $\mathcal{A}$ and every pair $(M_0, \beta)$, $M_0 \in [M_0^1,\ M_0^2]$, $\beta \in [\beta_1, \beta_2]$ we may compare the quality of classifications (7.34) and (7.35) and calculate the corresponding non-randomness and reliability functions.

We represent these functions by $\mathcal{Q}_\nu(M_0, \beta, \mathcal{A})$ and $\mathcal{P}_\ell(M_0, \beta, \mathcal{A})$ respectively. In the case of the reliability function in (7.2.0), instead of using $\nu(g^\star(\mathcal{O}), \mathcal{B}_0)$ in the real recognition procedure, we use instead the frequency $\nu(g(\mathcal{O}(M_0), \mathcal{B}_0(M_0)))$.

Actually, we do not know the real recognition procedure $g^\star$ in this case. At the same time, as is shown in [ ] under some natural assumptions we can assume that $\mathcal{P}_{g^\star}(\mathcal{O}(M_0), \ell) \leq \mathcal{P}(M_0, \beta, \mathcal{A})$ where $\mathcal{P}_{g^\star}(\mathcal{O}(M_0), \ell)$ is the reliability function defined in 7.2 by the formula (7.32). Thus, by minimizing $\mathcal{P}(M_0, \beta, \mathcal{A})$ we minimize the probability of the classification error.

Therefore, we have the function $\mathcal{Q}_\nu(M_0, \beta, \mathcal{A})$ and $\mathcal{P}(M_0, \beta, \mathcal{A})$ which characterize the quality of (7.34) for given $M_0 \in [M_0^1,\ M_0^2]$, $\beta \in [\beta_1, \beta_2]$ and $\mathcal{A}$. Let the algorithm $\mathcal{A}$ be fixed. Then, for the definition of the classification problem, it is natural to take $\overline{M_0}$ and $\overline{\beta}$ such that (7.34) gives the most non-random and reliable classification. In other words, we have to find the pairs $(M_0^{\mathcal{Q}}, \beta^{\mathcal{Q}})$ and $(M_0^{\mathcal{P}}, \beta^{\mathcal{P}})$ such that for given algorithm $\mathcal{A}$ we have:

$$\mathcal{Q}_\nu(M_0^{\mathcal{Q}}, \beta^{\mathcal{Q}}, \mathcal{A}) = \inf\left\{\mathcal{Q}_\nu(M_0, \beta, \mathcal{A}) :\ M_0 \in [M_0^1, M_0^2],\ \beta \in [\beta_1, \beta_2]\right\}. \tag{7.36}$$

$$\mathcal{P}_\ell(M_0^{\mathcal{P}}, \beta^{\mathcal{P}}, \mathcal{A}) = \inf\left\{\mathcal{P}_\ell(M_0, \beta, \mathcal{A}) :\ M_0 \in [M_0^1, M_0^2],\ \beta \in [\beta_1, \beta_2]\right\}. \tag{7.37}$$

Generally speaking, $(M_0^{\mathcal{Q}}, \beta^{\mathcal{Q}}) \neq (M_0^{\mathcal{P}}, \beta^{\mathcal{P}})$. Thus, we take one of them, or one of the mixed pairs $(M_0^{\mathcal{Q}}, \beta^{\mathcal{P}})$ or $(M_0^{\mathcal{P}}, \beta^{\mathcal{Q}})$.

Another approach is to minimize the functional

$$f(M_0, \beta, \mathcal{A}) = a\mathcal{Q}_\nu(M_0, \beta, \mathcal{A}) + b\mathcal{P}_\ell(M_0, \beta, \mathcal{A})$$

where $a$ and $b$ are the weights "responsible" for non-randomness and reliability in the specific problem under consideration.

## 7.4 Additional Arguments for Evaluating the Classification Reliability

Additional conditions for evaluating the reliability of a solution of the classification problem can be constructed using earthquake data that were not previously used in the formulation and refinement of the classification problem [82]. In the formulation of the classification problem, in order to form the learning material $\mathcal{B}_0$ and the control groups $\mathcal{B}_1, \cdots, \mathcal{B}_k$ we have used only the data on epicenters of earthquakes having $M \geq M_0$ and occuring in the time window $[T_0, t'']$ ($t''$ is the present moment of time) and when the earthquake parameters have been defined in a sufficiently reliable way. For this time interval, we know the dynamics of the problem as reflected in the classifications

$$\mathcal{W} = \mathcal{B}(t') \bigsqcup \mathcal{H}(t'); \quad t' \in [T_0, t"] . \tag{7.38}$$

Let we have a solution $g^\star(\mathcal{O}(t), p, \mathcal{A}) : \mathcal{W} \bigsqcup \mathcal{H}$, obtained $(t" - t)$-years ago. Then in the present moment of the time $t" > t$ we may have a number of epicenters of earthquakes with $M \geq M_0$ which has not been used in the recognition procedure. *Thus, the fact of clustering of those epicenters to the objects $w \in \mathcal{B}$ is obvious by a condition in favor of consideration of the solution of the classifiction problem as prediction.*

Less reliable condition of such type we obtain using known historical (non-instrumental) epicenters in the region. In other words the data about the earthquakes occured when $t' < T_0$. Because the magnitudes of historical earthquakes are defined rather approximately, it is natural to introduce the following carefull condition :

*The places of historical earthquakes with high macroseismic intensities should not appear in sufficiantly small vicinities of objects $w \in \mathcal{H}$.*

Let the locations of strong historical earthquakes occur to be near some groups of objects $\mathcal{B}_{k+1}, \cdots, \mathcal{B}_\ell \subset \mathcal{W}$. Then we may use these groups as additional control groups to $\mathcal{B}_1, \cdots, \mathcal{B}_k$ defined in the formulation of the classification problem (see 6.1). That means that for each $i = 1, \cdots, k, k+1, \cdots, \ell \;\; \exists w \in \mathcal{B}_i : \; w \in \mathcal{B}$.

Let $\mathcal{B}$ be the set of objects in a given problem for $M \geq M_0$ and $\mathcal{W}' \subset \mathcal{W}$ is it's subset which is included, at the same time, in another problem for $M \geq M_0'$. If $M_0 > M_0'$, then the following condition is obviously important to accept of the classification as prediction:

*Let $w \in \mathcal{W}'$. If $w \in \mathcal{B}$ in the problem with $M \geq M_0$ then $w \in \mathcal{B}$ in the problem $M \geq M_0'$.*

Considering our limit pattern recognition problem we assume that at the present moment of time we pose sufficient information to obtain the searched classification for $t \to \infty$. However, it may happen that we actualy have this information earlier in the moment of time $t' < t"$ (for example 10 years ago).

This allows us to formulate the following additional condition for verification of the classification reliability.

Let

$$g^{\star}(\mathcal{O}(t'),\ p,\ \mathcal{A}) : \mathcal{W} = \mathcal{B} \bigsqcup \mathcal{H} \tag{7.39}$$

be a solution of the classification problem. At the moment $t' < t"$ some objects $w_1, \cdots, w_s \in \mathcal{B}_0(t")$ are still unknown as the ones connected with epicenters of earthquakes with $M \geq M_0$. Thus, for the moment $t'$ we have the following learning material:

$$\mathcal{O}(t') = \left(\mathcal{B}_0(t") \setminus \{w_1,\ \cdots,\ w_s\},\quad \mathcal{H}_0(t) \bigcup \{w_1,\ \cdots,\ w_s\}\right). \tag{7.40}$$

If, as it is assumed, in the moment $t = t'$ we have sufficient information to resolve the problem, then the classification

$$g^{\star}(\mathcal{O}(t'),\ p,\ \mathcal{A}) : \mathcal{W} = \mathcal{B}_{t'} \bigsqcup \mathcal{H}_{t'} \tag{7.41}$$

should be equal to (7.39). If it is the case, then we may require the same equivalence in any moment of time $t : t' < t < t"$.

The fulfilment of the condition (7.41) obviously speaks in favor of the prediction reliability. However it cannot, strictly speaking, be considered as a necessary condition. Really, we do not know *a priori* a moment of time $t' < t"$ such as the information $\mathcal{O}(t')$ is sufficient for learning.

Let the learning material $\mathcal{O}(t)$ gives us a superfluous information. That means that $\exists \mathcal{Q} \subset \mathcal{B}_0(t)$ such as there is a procedure $g^{\star}(\mathcal{O}_{\mathcal{Q}}(t),\ p,\ \mathcal{A})$, where $\mathcal{O}_{\mathcal{Q}}(t) = (\mathcal{B}_0(t) \setminus \mathcal{Q},\ \mathcal{H}_0(t) \bigcup \mathcal{Q})$ which gives the classification equal to (7.40). Such situation obviously speaks for reliability of the classification. If such $\mathcal{Q}$ does not exist, then we may require that the clasification

$$g^{\star}(\mathcal{O}_{\mathcal{Q}}(t),\ p,\ \mathcal{A}) : \mathcal{W} = \mathcal{B}_{\mathcal{Q}} \bigsqcup \mathcal{H}_{\mathcal{Q}} \tag{7.42}$$

is sufficiently close to (7.40) and the control set which consists of the objects $w \in \mathcal{B} \bigcap \mathcal{Q}$ is classified such as $\mathcal{Q} \subset \mathcal{B}_{\mathcal{Q}}$.

It may occur that the learning material $\mathcal{O}(t)$ is so superfluous that the classifications

$$g^{\star}(\mathcal{O}(\mathcal{Q}),\ p,\ \mathcal{A}) : \mathcal{W} = \mathcal{B}(\mathcal{Q}) \bigsqcup \mathcal{H}(\mathcal{Q}) \tag{7.43}$$

where $\mathcal{O}(\mathcal{Q}) = (\mathcal{Q},\ \mathcal{H}_0(t))$ and the classification (7.42) are equal. Furthermore to make the last condition less strong we may require that the classifications (7.40), (7.42) and (7.43) are sufficiently close to each other. We also may allow to change the subspaces $\rho(\Omega_n)$ in (7.40), (7.42) and (7.43).

Actually, $\mathcal{Q}$ may be choosen in different ways

In particular, the above conditions may take place for a "compact" set of objects $\mathcal{Q}$, located close to each other in some part of the region.

Another important operation to evaluate the results reliability is a transfer of high seismicity criteria to another region.

Let we have a solution of our classification problem. We formulate criteria of the fact that an object $w$ is included into the class $\mathcal{B}$ in terms of initialy defined parameters $x_1(w),\ \cdots,\ x_\ell(w)$. Such criteria we call as criteria of high $(M \geq M_0)$ seismicity.

We assume that $v \in \mathcal{V}$ are the objects of recognition in another (new) seismically active region which are constructed in the same way that the objects $w \in \mathcal{W}$. We symbolize by $\mathcal{V}_0 \subset \mathcal{V}$ subset of objects to which known epicenters of strong earthquakes are clustered. We assume also that the objects $v \in \mathcal{V}$ are represented by vectors in the same suspace of parameters $p(\Omega_n)$, in which the initial classification

$$g^{\star}(\mathcal{O},\ p,\ \mathcal{K}) : \mathcal{W} = \mathcal{B} \bigsqcup \mathcal{H} \tag{7.44}$$

has been obtained.

Herein, is a concrete description of the high seismicity criteria in case of some widely used VSF algorithms (see Chapter 3).

Algorithm VEF-0 Criteria of high seismicity is the kernel $z$ of the class $\mathcal{B}$. The application of the criteria to the region $\mathcal{V}$ is the calculation of the distance $\rho(z, v)$ for every $v \in \mathcal{V}$.

Algorithms VEF-1 and VEF-2 Criteria of high seismicity is the kernel $z$ and the vector of weights $\sigma = (\sigma_1,\ \cdots,\ \sigma_\ell)$ ( $\ell = \dim p(\Omega_n)$). An application of the criteria to the region $\mathcal{V}$ is the calculation of the distance $\rho^{\sigma}(z, v)$ for every $v \in \mathcal{V}$.

Algorithm $\mathcal{B}$ Criteria of high seismicity in this case is the set

$\{\mathcal{P}_1(\mathcal{B}_0),\ \cdots,\ \mathcal{P}_\ell(\mathcal{B}_0),\ \mathcal{P}_1(\mathcal{H}_0),\ \cdots,\ \mathcal{P}_\ell(\mathcal{H}_0)\}$. An application of the criteria to the region $\mathcal{V}$ is the calculation of *a posteriori* probabilities

$$P\{v \in \mathcal{B} \mid v = (v(1),\ \cdots,\ v(\ell))\}$$

Algorithm $\mathcal{K}$ Criteria of high seismicity in this case are the sets $\mathcal{R}_{\mathcal{K}_\mathcal{B},\widetilde{\mathcal{K}_\mathcal{B}}}(\mathcal{B}_0)$ and $\mathcal{R}_{\mathcal{K}_\mathcal{H},\widetilde{\mathcal{K}_\mathcal{H}}}(\mathcal{H}_0)$ of the characteristics features of the classes $\mathcal{B}$ and $\mathcal{H}$. An operation of application of these criteria to the region $\mathcal{V}$ is the calculation $\forall v \in \mathcal{V}$ of the difference $\Delta(v) = \mathcal{N}_\mathcal{B}(v) - \mathcal{N}_\mathcal{H}(v)$ where $\mathcal{N}_\mathcal{B}(v)$ and $\mathcal{N}_\mathcal{H}(v)$ are the numbers of features from $\mathcal{R}_{\mathcal{K}_\mathcal{B},\widetilde{\mathcal{K}_\mathcal{B}}}(\mathcal{B}_0)$ and $\mathcal{R}_{\mathcal{K}_\mathcal{H},\widetilde{\mathcal{K}_\mathcal{H}}}(\mathcal{H}_0)$ correspondingly, which posses the vector $\Phi(v)$.
Therefore, without actual recognition procedure we may obtain the following classification of the set V :
We symbolize

$$\mathcal{B}^\mathcal{V}(\Delta) = \left\{ v \in \mathcal{B}: \begin{cases} \rho(z,v) \leq \Delta, & \text{if } A = VEF-0 \\ \rho^\sigma(z,v) \leq \Delta, & \text{if } A = VEF-1 \text{ or} \\ & \quad A = VEF-2 \\ P(v) \geq \Delta, & \text{if } A = \mathcal{B} \\ \Delta(v) \geq \Delta, & \text{if } A = \mathcal{K} \end{cases} \right.$$

Thus, to every $\Delta$ we correspond the classification $\mathcal{V} = \mathcal{B}^\mathcal{V}(\Delta) \bigsqcup \mathcal{H}^\mathcal{V}(\Delta)$, where $\mathcal{H}^\mathcal{V}(\Delta) = \mathcal{V} \setminus \mathcal{B}^\mathcal{V}(\Delta)$. Having in the $\mathcal{V}$-region the epicenters of strong earthquakes ($M \geq M_0$) we may (as in 6.1) define the control groups of objects $\mathcal{V}_1,\ \cdots,\ \mathcal{V}_k$ and give an evaluation $\beta_\mathcal{V}$ of an *a priori* probability of an object $v \in \mathcal{B}$ to be included into the class $\mathcal{B}^\mathcal{V}$.
Based on all of this we formulate the following:

**Definitions**: The transfer of the high seismicity criteria is considered as successful if $\exists\Delta$ such as classification $\mathcal{V} = \mathcal{B}^\mathcal{V}(\Delta) \bigsqcup \mathcal{H}^\mathcal{V}(\Delta)$ satisfies to the following conditions:

1. $\mathcal{B}^\mathcal{V}(\Delta) \supset \mathcal{V}_0$

2. $\forall j = 1,\ \cdots,\ k$ there is an object $v \in \mathcal{B}^\mathcal{V}(\Delta)$ such as $v \in \mathcal{V}_j$.

3. $\mid \mathcal{B}^\mathcal{V}(\Delta) \mid \leq \beta_\mathcal{V} \mid \mathcal{V} \mid$.

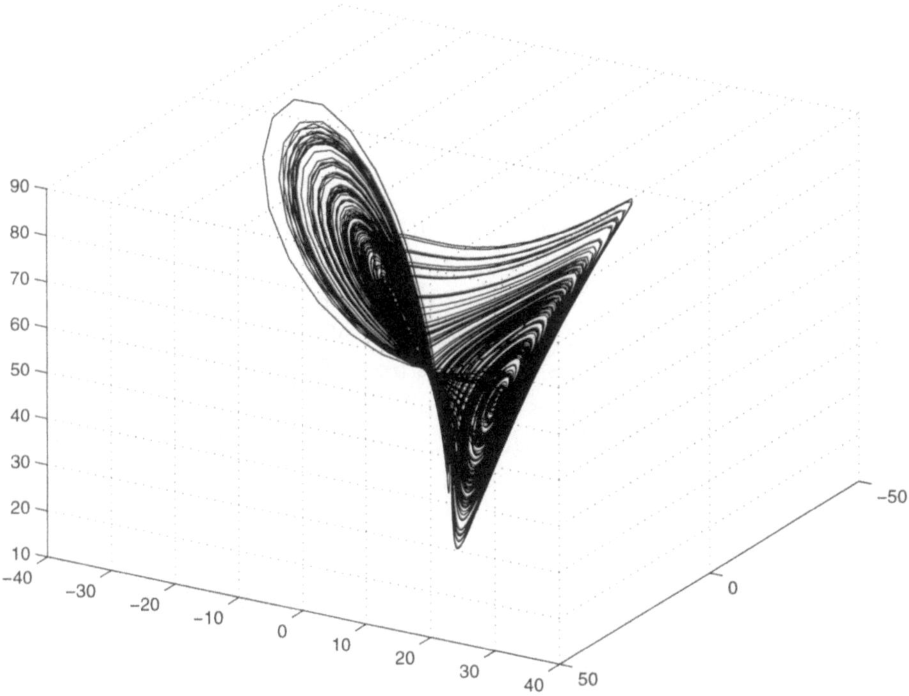


The Lorenz Attractor – The parameters are the same as in page 4, but the perspective representation is different. It shows in the phase space shows 10,000 points discrete phase trajectories (after David AUBERT, 1997).

# Part III

# Dynamic Systems

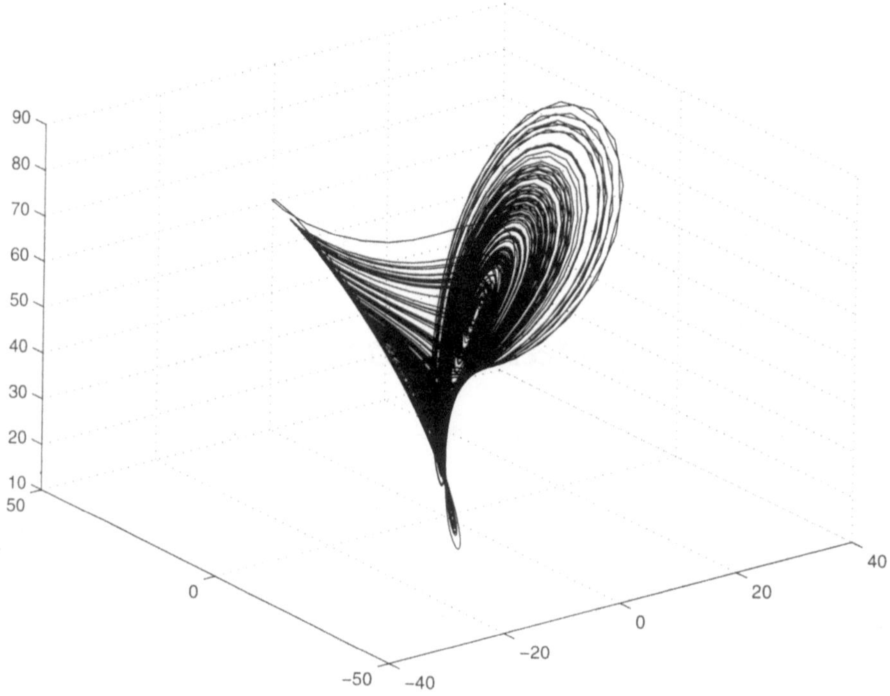


The Lorenz Attractor – The parameters are the same as in page 4, but the perspective representation is different. It shows in the phase space shows 10,000 points discrete phase trajectories (after David AUBERT, 1997).

# Chapter 8

# Basic Definitions and Facts

This part is devoted, in some sense, to alternative approach to the studies of large data sets of geophysical data. On one hand, pattern recognition approach described in the previous part of the book deals with a search of deterministic and statistic roots in the sets of data under investigation to discover some expected features of the actual geophysical process. On the other hand, the dynamic systems approach introduced in this part of the book deals directly with modeling of large sets of geophysical data on the basis of differential equations techniques. In fact, the ideal situation would be when both approaches are applicable to the same data sets. In this case, an important part of the study is comparison of the results obtained by pattern recognition analysis and dynamic systems behaviour studies. Positive results of these comparisons give the basis to construct actual models of the studied geophysical phenomena. Both approaches introduce new non-linear techniques and tools in data processing of long time series.

## 8.1 Measure and Dimension

As we shall see, the study of dynamic systems leads to the general notion of attractor, a geometrical object in the phase space which does not necessarily have an integer dimension. This object or set may be fractal, and this is the most frequent case. To understand better such strange attractors, which will be introduced in section 8.3, let us begin by defining measure and dimension.

### 8.1.1 Hausdorff measure

Let $\mathcal{V}$ be a *non empty subset* of a *n-dimensional Euclidean space* $\Re^n$. We define its *diameter* as

$$|\mathcal{V}| = \sup\{|x-y|: \quad x,\ y\ \in\ \mathcal{V}\} \tag{8.1}$$

which corresponds to the greatest distance between all the pairs of points in $\mathcal{V}$. Let us now introduce a collection of sets of diameter $\delta$ that cover a set $\mathcal{A}$

$$\mathcal{A} =\subset \bigcup_{i=1}^{\infty} \mathcal{V}_i\,, \tag{8.2}$$

such that, $0 < |\mathcal{V}_i| \leq \delta$, $i = 1, \ldots, \infty$. We say that $\{\mathcal{V}_i\}$ is a $\delta-$cover of $\mathcal{A}$.

Figure 8.1: *Notion of covering.* A set $\mathcal{A}$ and two possible covers by elements $\delta$ (after Falconer, 1990).

If $\mathcal{A}$ is a subset of $\Re^n$ and $d \in \Re$ a non-negative number, then for any $\delta > 0$ we define

$$\mathcal{H}^d_\delta(\mathcal{A}) = \inf\left\{\sum_{i=1}^{\infty} |\mathcal{V}_i|^d: \quad \{\mathcal{V}_i\} \text{ is a } \delta - \text{cover of } \mathcal{A}\right\}. \tag{8.3}$$

Thus $\mathcal{H}^d_\delta(\mathcal{A})$ is constructed by taking all covers of $\mathcal{A}$ by sets of diameter not exceeding $\delta$ and by minimizing the sum of the $d$th power of the diameters. The infimum increases and approaches a limit when $\delta \to 0$:

$$\mathcal{H}^d(\mathcal{A}) = \lim_{\delta\to 0} \mathcal{H}^d_\delta(\mathcal{A}). \tag{8.4}$$

**Definition**: The value $\mathcal{H}^d(\mathcal{A})$ is called the *d−dimensional Hausdorff measure* of the set $\mathcal{A}$.

### 8.1.2 Hausdorff dimension

Having defined the Hausdorff measure, let us define the Hausdorff dimension. From equation 8.3, it is evident that, for a given set $\mathcal{A}$ and $\delta < 1$, $\mathcal{H}^d_\delta(\mathcal{A})$ does not increase with increase of $d$. As is clear from 8.4, this is true also for $\mathcal{H}^d(\mathcal{A})$.

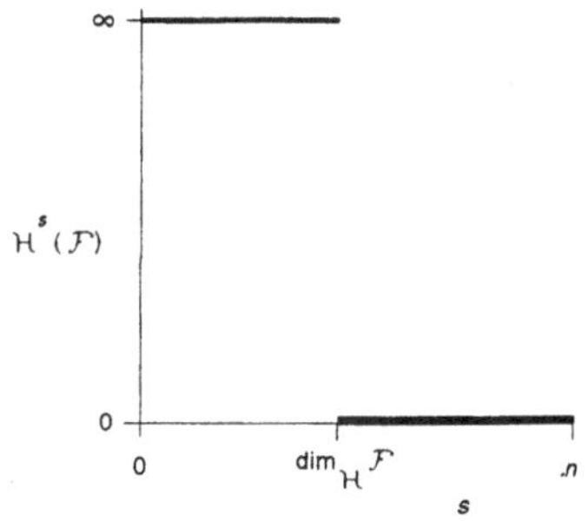


Figure 8.2: *The Hausdorff dimension.* $\mathcal{H}^d(\mathcal{A})$ against $d$ for a set $\mathcal{A}$. The Hausdorff dimension is the $d$ value where the measure jumps from $\infty$ to $0$ (after Falconer, 1990).

Thus, if $d' > d$ and $\{\mathcal{V}_i\}$ is a $\delta$-cover of $\mathcal{A}$ we get,

$$\sum_i | \mathcal{V}_i |^{d'} \leq \sum_i | \mathcal{V}_i |^{d} . \tag{8.5}$$

Taking the infimum we obtain,

$$\mathcal{H}^{d'}_\delta(\mathcal{A}) \leq \delta^{d'-d} \mathcal{H}^d_\delta(\mathcal{A}) . \tag{8.6}$$

If $\delta \rightarrow 0$ we see that when $\mathcal{H}^d(\mathcal{A}) < \infty$, then $\mathcal{H}^{d'}(\mathcal{A}) = 0$ for $d' > d$. Thus, the graph of $\mathcal{H}^d(\mathcal{A})$ as a function of $\delta$ shows that there exists a critical value of $d$ at which $\mathcal{H}^d(\mathcal{A})$ jumps from $\infty$ to $0$.
This critical value is called *the Hausdorff dimension of* $\mathcal{A}$.

To better understand this definition, let us give a classical though very simple example. Let $\mathcal{A}$ be an infinitely flat disc of unit radius in $\Re^3$. It is clear that the measure $\mathcal{H}^1(\mathcal{A}) = \infty$ (of course, the number of segments on a one dimensional line necessary to cover the whole disc is infinite). Now, if we consider in a three dimensional space the number of elementary volumes necessary to cover the flat disc we get a measure $\mathcal{H}^3(\mathcal{A}) = 0$. Thus the finite value of the measure of the disc will be obtained when $\dim_H(\mathcal{A}) = 2$ and $\mathcal{H}^d(\mathcal{A}) = \infty$ if $d < 2$ and $\mathcal{H}^d(\mathcal{A}) = 0$ when $d > 2$.

## 8.2 Discrete Dynamic Systems

To begin, let us recall some basic set theory notions.
Let $\mathcal{X}$ and $\mathcal{Y}$ be any sets. A **transformation (mapping, function)** $f$ from $\mathcal{X}$ to $\mathcal{Y}$ is a formula that associates a point $f(x)$ of set $\mathcal{Y}$ with each point $x$ of set $\mathcal{X}$.
We write $f : \mathcal{X} \to \mathcal{Y}$. $\mathcal{X}$ is the **domain** of $f$ and $\mathcal{Y}$ is the **codomain**. For any subset $\mathcal{A}$ of $\mathcal{X}$, $f(\mathcal{A}) = \{f(x) : x \in \mathcal{A}\}$ is the **image** of $\mathcal{A}$. For any subset $\mathcal{B}$ of $\mathcal{Y}$, $f^{-1}(\mathcal{B}) = \{x \in \mathcal{X} : f(x) \in \mathcal{B}\}$ is the **inverse-image** or **pre-image** of $\mathcal{B}$. The function $f : \mathcal{X} \to \mathcal{Y}$ is called an **injection** function when $f(x) \neq f(y)$ whenever $x \neq y$. It is called a **surjection** or an **onto** function when, for every $y \in \mathcal{Y}$, there is an element $x \in \mathcal{X}$ such as $f(x) = y$. We call a **bijection** or **one-to-one correspondance** between $\mathcal{X}$ and $\mathcal{Y}$ a function that is both an injection and a surjection.
If $f : \mathcal{X} \to \mathcal{Y}$ is a bijection, we may define the **inverse function** $f^{-1} : \mathcal{Y} \to \mathcal{X}$ by taking $f^{-1}(y)$ as the unique element of $\mathcal{X}$ so that $f(x) = y$. It follows, $f^{-1}(f(x)) = x$ for $x \in \mathcal{X}$ and $f(f^{-1}(y)) = y$ for $y \in \mathcal{Y}$.

Let us now define a **discrete dynamic system**.
Let $\mathcal{X}$ be a subset of $\Re^n$, and let $f : \mathcal{X} \to \mathcal{X}$ be a continuous mapping on a **metric space** $(\mathcal{X}, d)$. It is denoted by $\{\mathcal{X}; f\}$. $f^k$ denotes the $k$th iterate of $f$, so that $f^0(x) = x$, $f^1(x) = f(x)$, $f^2(x) = f(f(x))$, etc. An iterative scheme $\{f^k\}$ is called a **discrete dynamic system**. Typically $x$, $f(x)$, $f^2(x)$, ..., are the values of some quantity at times $0, 1, 2, \ldots$ and the value $k+1$ is given in terms of the value at $k$ by the function $f$.

The **orbit** of a point $x \in \mathcal{X}$ is the sequence $\{f^{\circ n}(x)\}_{n=0}^{\infty}$. The exponent $\circ n$ means followed by $n$ times, according to Barnsley's (1988) notation [16].

A **periodic point** of $f$ for a given dynamic system $\{\mathcal{X}; f\}$ is a point $x \in \mathcal{X}$ such that $f^{\circ n}(x) = x$ for some $n \in \{1, 2, 3, \ldots\}$. If $x$ is a periodic point of $f$ then an integer $n \in \{1, 2, 3, \ldots\}$ such that $f^{\circ n}(x) = x$ is called a **period of x**. The minimum among such integers is called the **minimal period** of the periodic point $x$. The orbit of a periodic point of $f$ is called a **cycle** of $f$. The minimal period of a cycle is the number of distinct points which it contains. A period of a cycle of $f$ is a period of a point in the cycle.

Let us look into the behaviour of the sequences of iterates, and orbits for various initial points, as $k$ becomes large. Sometimes $f^k$ converges to a fixed point when $k \to \infty$ for any initial $x$ (i.e., $f(x) = \cos x$). Sometimes it converges alternativaly to a fixed point or to period-$p$ points; sometimes it moves about at random, remaining in the proximity of a certain set which may be a fractal. These points or subsets are called **attractors** and when they are fractal sets they are **fractal attractors** or **strange attractors**. More precisely, we call a subset $\mathcal{A}$ of $\mathcal{X}$ an attractor for $f$ if $\mathcal{A}$ is a closed set so that an invariant under $f$ such as the distance from $f^k(x)$ to $\mathcal{A}$ converges to zero as $k \to \infty$ for all $x$ in an open set $\mathcal{B}$ containing $\mathcal{A}$. The set $\mathcal{B}$ is the **basin of attraction** of $\mathcal{A}$. Similarly, a closed invariant set $\mathcal{R}$ from which all close points not in $\mathcal{R}$ are iterated away is called a **repeller**.

## 8.3 Continuous Dynamic Systems

We may define continuous dynamic systems as the limit of discrete dynamic systems as the time interval is allowed to tend to zero. Then we get differential equations in the usual way.
Let $\mathcal{X}$ be a domain in $\Re^n$ and $f: \mathcal{X} \to \Re^n$ be a smooth function.
The differential equation

$$\dot{x}(t) = dx/dt = f(x) \tag{8.7}$$

has a family of trajectories which fill $\mathcal{X}$.
As with discrete dynamic systems, continuous dynamic systems give rise to attractors and repellers. A closed subset $\mathcal{A}$ of $\mathcal{X}$ may be termed an attractor with basin of attraction $\mathcal{B}$ containing $\mathcal{A}$ if, for all initial points $x(t_0)$ in the open set $\mathcal{B}$, the trajectory $x(t)$ through $x(t_0)$ approaches $\mathcal{A}$ as $t \to \infty$.
When $\mathcal{X}$ is a plane domain, the range of attractors for a continuous system is rather limited. They can only be isolated points $x$ for which $f(x) = 0$ in equation 6.7, or closed loops. The **Poincaré-Bendixon theorem** claims that no more complicated attractors may occur, so to find dynamic systems with fractal attractors we need to look at systems in 3 or more dimensions. When we have linear differential equations in 6.7, we may solve them completely and the solutions involve simple periodic or exponential terms; however, non-linear terms can lead to very complicated trajectories. Now, if one considers non-linear differential equations in a 3-dimensional domain, one standard approach is to reduce a 3-dimensional continuous sytem to a 2-dimensional discrete system by looking at plane cross sections (see Poincaré cross-section).

## 8.4 Representation and Study of Dynamic Systems

Before defining phase space (which is very convenient for both representation and study of dynamic systems) let us give some preliminary definitions.

### Degrees of freedom.

To determine correctly the position of an $N$ point system in any space, it is necessary to give $N$ vectors, that is to say, $3N$ coordinates, for example, in a 3 dimensional space. The number of independent values necessary to determine uniquely the position of the system is called its **number of degrees of freedom** (Landau and Lifchitz, 1967) [120].
If the values $q_1, \ldots, q_s$ define completely the position of a system ($s$ degrees of freedom), they are called generalized coordinates and the derivatives $\dot{q}_i$ their generalized velocities. Knowing both coordinates and velocities, we may completely determine the state of the system and theoretically we may predict its future state.

### Phase space and phase trajectory.

Still following Landau and Lifchitz (1967) [120], we consider a macroscopic mechanical system with $s$ degrees of freedom. The position of each point of the system in the physical space is characterized by $s$ coordinates which we call by $q_i$, $i = 1, \ldots, s$, and corresponding velocities $\dot{q}_i$. The dynamic state of the system, at any time, is completely determined by the simultaneous values of the $s$ coordinates $q_i$ and their corresponding velocities $\dot{q}_i$. So any dynamic system has a $2n$ dimension phase space where $n$ is the number of degrees of freedom.
Each point in the phase space corresponding to given values of the system coordinates $q_i$ and velocities $\dot{q}_i$ represents a given state of the system.
In other words, a state of the system, at any moment, corresponds to a point in the phase space. When the system state varies over time, the corresponding point in the phase space moves along a curve which is called **phase curve** or **phase trajectory**.
In his book "Catastrophe Theory", Arnol'd (1984) [6] defines the previous items in terms of evolutionary process: " An *evolutionary process* is described mathematically by a vector field in phase space. A point of phase space defines the *state* of the system. The vector at this point indicates the velocity of the change of state."

**Notion of flow in the phase space**

From a global point of view, the evolution of many systems can be suitably described by a set of $n$ first-order differential equations like:

$$\frac{d}{dt}\overrightarrow{X(t)} = F(\overrightarrow{X},\ t)\,. \tag{8.8}$$

$\overrightarrow{X}$ is a vector in $\Re^n$ (phase space), and $F$ is a function of $\overrightarrow{X}$ and time. Such a differential equation system may be called **flow in** $\Re^n$.

- When $F$ does not depend on time, the flow is called **autonomous flow.**
- In the opposite case it is a **non-autonomous flow**.

Generally, the equation 8.8 has only an analytical solution in a few well defined particular situations where the flow is integrable, and an examination of each solution on the phase trajectory in its phase space needs to be studied. This last procedure may be simplified using Poincaré technique.

## 8.4.1 Iterative scheme as a tool in dynamic system study

In an abstract context the term "dynamic system" denotes a system of equations of motion. A first formulation is a system of differential equations with real variables which is given as,

$$dx_i/dt = f_i(x_1, x_2, \ldots, x_m)\,, \tag{8.9}$$

and

$$dx_i/dt = f_i(x_1, x_2, \ldots, x_m, t), \quad i = 1, 2, \ldots, m\,, \tag{8.10}$$

where $t$ is the time. The first equation is an **autonomous equation** while the second is a **non autonomous**, $m$ is the order of the equations, and $t_0$ and $x_i(t_0) = x_{i0}$ define the initial conditions. The solution $x_i = x_i(t, x_{i0}, t_0)$ varies continuously with time and it is given in the coordinate space $x_1, x_2, \ldots, x_m$, called **phase space**, by a curve, the parametric equation of which is $x_i = x_i(t, x_{i0}, t_0)$ which crosses the point $x_{i0}$, $i = 1, 2, \ldots, m$. This curve is the **phase trajectory**.
In the case where $t$ doesn't vary continuously, but, instead, according to an integer series $n$ called **discrete time**, we may write

$$x_{n+1}^{(i)} = f^{(i)}\left(x_n^{(1)},\ x_n^{(2)},\ \ldots,\ x_n^{(p)}\right), \tag{8.11}$$

or

$$x_{n+1}^{(i)} = f^{(i)}\left(x_n^{(1)},\ x_n^{(2)},\ \ldots,\ x_n^{(p)},\ n\right), \quad i = 1,\ \ldots,\ p; \quad n = 1,\ 2,\ \ldots . \tag{8.12}$$

8.11 is an **autonomous equation**, while 8.12 a **non- autonomous equation.**

Owing to some conditions of existence and uniqueness of solution, corresponding to an initial point $x_{i0}$, $i = 1,\ 2,\ \ldots,\ p$, in the phase space, this solution is given under the form of a series of points $x_n$, $n = 0,\ 1,\ 2,\ \ldots,$ called a **discrete phase trajectory.**

The equations 8.11 and 8.12 are called **iterative** and can be also designed as **punctual transformations or mappings**, because they can be interpreted as the transformation of the point $M_n(x_n^{(1)},\ \ldots,\ x_n^{(p)})$ to the point $M_{n+1}(x_{n+1}^{(1)},\ \ldots,\ x_{n+1}^{(p)})$ in the phase space.

It is well known that many problems in Celestial Mechanics or natural systems lead to formulation in discrete dynamic systems such as 8.11 and 8.12. Thanks to the system 8.11, many complex dynamic behaviors resulting from strong non-linearity were observed and described. Such behaviors are linked to the notion of **chaos**.

Ever since Poincaré's classic work, it is well known that certain problems dealing with oscillation theory, in the form of differential equations may be studied by iterative schemes of lower order. Thus, the solutions of the autonomous differential equation 8.9 with $m = 2$ (order 2), may be defined by an autonomous iterative equation 8.11 with $p = 1$, (order 1). This is the **limit cycle theory** of Poincaré.

The same holds true for autonomous differential equations of order $m = 4$: if we know a first integral which reduces the system to order $m = 3$, and then if we apply 8.11, it is possible to obtain an order $p = 2$ iterative sheme using a Poincaré section.

### Formalism in iterative schemes

The behaviour of many dynamic systems may be described by an order (dimension) $p$ iterative process, either by an explicit system of equations, or by an implicit formula,

$$F_i\left[x_{n+1}^{(1)},\ x_{n+1}^{(2)},\ \ldots,\ x_{n+1}^{(p)},\quad x_n^{(1)},\ x_n^{(2)},\ \ldots,\ x_n^{(p)}\right]=0,\quad i=1,\ \ldots,\ p. \tag{8.13}$$

in the autonomous case, and with $F_i$ depending on $n$ in the non-autonomous case.

The evolution of a dynamic process may be described by 8.11 or 8.13 in two possible cases,

- Only measures, or observations are sampled in time domain, so that the information is discrete.
- Measures or observations are continuous in time domain and the process is described by an $m$ order differential equation. From this equation, formulations such as 8.11 and 8.13 allow us to diminish the dimension of the initial problem $p < m$, and thus to simplify the study of the solutions (Poincaré assumption).

### Iterative scheme, punctual transformations

In a vectorial form 8.11 and 8.12 may by written as,

$$X_{n+1}=F(X_n),\qquad X_{n+1}=F(X_n,\ n)\,, \tag{8.14}$$

where $X$ is a vector in an euclidian space with a finite number $m$ of dimensions (order $= m$), $F$ is a vectorial function of $X$ which may be an analytic function or piecewise continuous. The first of equations 8.14 may be written $\overline{X}=F(X)$ to represent a punctual transformation (also called substitution) and will be designated as $T$. The transformation means that a point $M$ extremity of vector $X$ is transformed to a point $\overline{M}$ extremity of vector $\overline{X}$,

$$\overline{M}=TM\quad \text{or } X_{n+1}=TX_n \tag{8.15}$$

When the initial point $M_0$ $(n=0)$ is given, equations 8.14 generate from $M_0$ a series of points $M_1,\ M_2,\ \ldots,\ M_n,\ M_{n+1},\ \ldots$ which are solutions of equation 8.14. These solutions represent a **discrete trajectory** in an $m$-dimensional phase space, or iterative series. The point $M_n$ is called the **order- $n$ iterate** of $M_0$.

Given an initial point $M_0$, it is iterated to the rank $r$, using equation 8.14. The $r$ points may be considered a **mapping** of $r$ successive transformations $T$, and the point $M_r$ may be considered as the transform point of $M_0$ by the transformation $T^r$ :

$$M_r = TM_{r-1} = T\,[T\,(\ldots(TM_0))] = T^r M_0\,. \qquad (8.16)$$

When using the second equation of 8.14 (non-autonomous equation) the point $M_r$ may be considered a transform of $M_0$ after $r$ punctual transformations $T_0,\ T_1,\ \ldots,\ T_{r-1}$:

$$M_r = T_{r-1}M_{r-1} = T_{r-1}\cdot T_{r-2}\ldots T_1\cdot T_0\;M_0\,. \qquad (8.17)$$

In equation 8.16 $M_r$ is the rank $r$ consequent of $M_0$, and $M_0$ is the rank $r$ antecedent of $M_r$.
If in 8.14 $F$ is continuous and differentiable and if, in the whole domain of definition of $F$, any point has one and only one antecedent, the transformation $T$ is called a **diffeomorphism**. If $T$ is such that, a point may have a non unique antecedent, $T$ is called an **endomorphism**.

### Iterative process (punctual transformation) and differential equations

As has been shown previously, an iterative process which is associated to a differential equation has, as its main purpose the reduction of the dimension of the problem. This is designated a **Poincaré section method**. Let us suppose that $m = 3$ in an autonomous differential equation 8.9. The method can be explained as in figure 8.3.
The section, here, is the plane $(P)$ to which the coordinate axes $x_n^{(1)},\ x_n^{(2)}$ are associated. From the initial position $M_0$ of the phase space, the phase trajectory intersects $(P)$ at $M_1,\ M_2,\ \ldots$. Two successive intersections are linked by a punctual bidimensional autonomous transformation, $M_{n+1} = TM_n$.
Any point $M$ which satisfies $M = TM$ is called **a fixed point of T** and corresponds to a periodic solution.
For the non-autonomous differential equation 8.10,

$$dx_i/dt = f_i(x_1,\ x_2,\ \ldots,\ x_m,\quad t),\ f_i$$

is a periodic function of $t$, of period $\tau$. For $m = 2$ the method of Poincaré section may be applied (figure 8.4). The solution is given in a space $(x_1,\ x_2,\ t)$, for an initial position $M_0$ $(t = 0)$ by a curve which intersects the planes $t = n\tau$, at points $M_n(x_n^{(1)},\ x_n^{(2)})$, $n = 0,\ 1,\ 2,\ \ldots$ Two successive intersections permit to define the two-dimensional autonomous transformation $M_{n+1} = TM_n$.

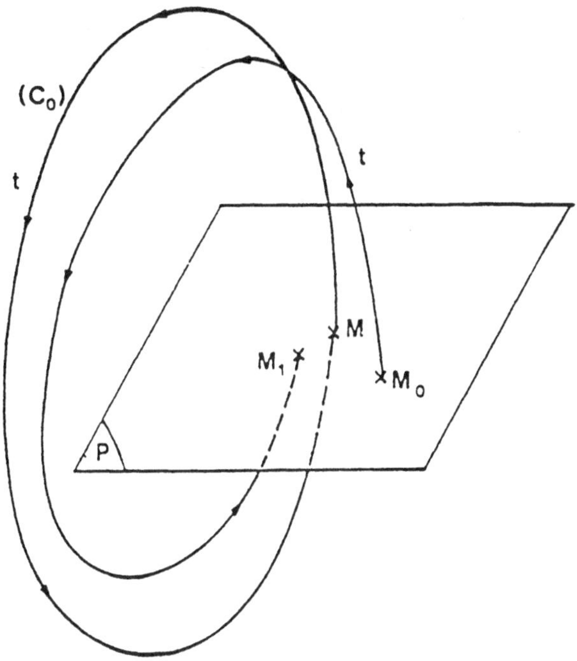


Figure 8.3: *Poincaré section in a 3-dimensional phase space.* The surface of a Poincaré section is the plane $P$ which is referenced to a coordinates axis $x_n^{(1)}$, $x_n^{(2)}$. The initial position $M_0$ of the phase trajectory intersects $P$ in $M_1$, $M_2$, .... M is a **fixed point** of transformation T (defined by equation 8.15) and corresponds to a periodic solution (after Mira, 1987).

A fixed point of $T$ corresponds either to an equilibrium position of equation 8.10 or to a periodic solution of period $\tau$.
Now, $k$ points which verify $M = T^k M$; $M \neq T^l M$, $l < m$, and resulting from one another by applying $T$, constitute an **order $k$ cycle**, which corresponds to a subharmonic oscillation of 8.10 having period $k\tau$.
One considers that by using $T$, there is a decrease of the effective dimension $m$ of the initial problem, because $T$, which has dimension $p = m$, is autonomous.

### Order-one linear iterative process, real variable

Let us consider an order-one linear recursive process

$$y_{n+1} = a + by_n\,, \quad a,\ b \text{ real}, \quad -\infty < y < +\infty\,. \tag{8.18}$$

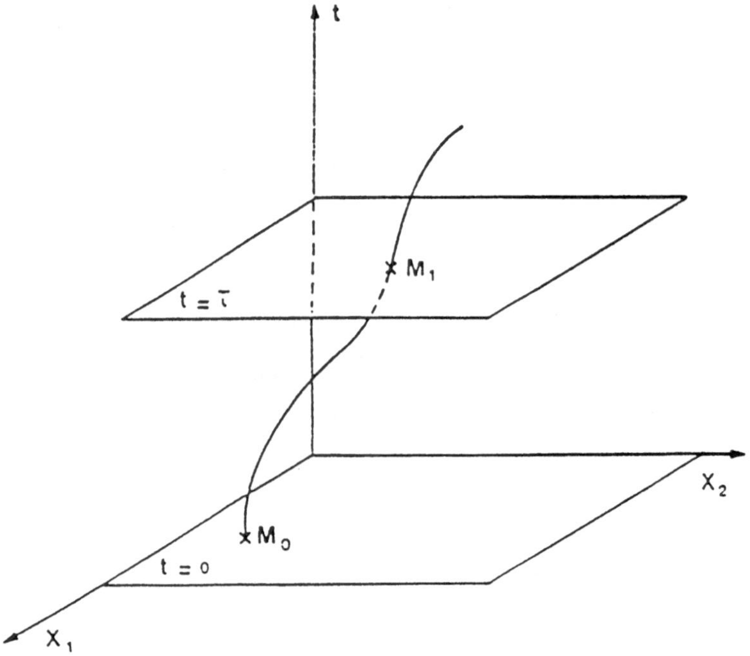


Figure 8.4: *A 2-dimensional autonomous transformation.* A three dimensional space $(X_1, X_2, t)$ and an autonomous transformation $T$ with a periodic solution of period $\tau$ (after Mira, 1987).

The point $\overline{y}$ that is a solution of

$$y_n = y_{n+1} = \overline{y}, \quad \overline{y} = a/(1-b)\,, \tag{8.19}$$

is called a **fixed point**. After the substitution, $y = x + \overline{y}$, we get,

$$x_{n+1} = Sx_n, \quad S = b \tag{8.20}$$

so that, for any initial condition $x_0$,

$$x_n = S^n x_0\,. \tag{8.21}$$

The origin O, $(x = 0)$ is the only fixed point and its nature depends on the real number $S$ called the **multiplicator** or **eigenvalue** of the fixed point $x = 0$.
If $\mid S \mid < 1$ for any $x_0$, the iterative series tends to $x = 0$. This point is called **attractive** or **asymptotically stable**. If $\mid S \mid > 1$, any point $x_0$

close to O gives a series of points, the distance to O of which increases with $n$. Such a point is a **repeller** or **instable point**.
The nature of O is also dependant on the sign of $S$. When $S$ is positive, the series $x_n$ always has the same sign as $x_0$. O is then called a **type 1 point**. When $S$ is negative, the sign of $x_n$ depends on the parity of $n$. The series oscillates around O. O is then called a **type 2 point**.
When O is attractive, the attraction basin of O is the set of points $x_0$ which form a recurent series tending to O. When O is a repeller the domain of stability of the infinite point is the $x$ axis, the frontier point of which is $x = 0$.
Given an initial point $x_0$, the point $x_1$ (given by $x_{n+1} = Sx_n$) is the consequent of $x_0$.
We may write $x_n = S^{-1}x_{n+1}$ which defines the inverse iterative series where $x_{-1}$ the rank one antecedent of $x_0$ and is obtained by the inverse transformation $T^{-1}$. In the linear case, $T^{-1}$ is uniform so that $x_0$ always has one and only one rank $n$ antecedent. The point $x_{-n}$ obtained from $x_0$ is the rank $n$ antecedent.

### Some generalities on nonlinear order-one iterative processes

**Complexity of the solution** Let us consider the iterative process or punctual transformation $T$ under the explicit form,

$$x_{n+1} = f(x_n), \quad -\infty < x < +\infty\,, \tag{8.22}$$

where $f(x)$ is a uniform function of the real variable $x$ and which cannot be written in the lineär form 8.18 by any substitution of variables.

As the class of such functions is very large, we may obtain a large set of properties of solutions different from those obtained in 8.18. If $f(x)$ is not continuously differentiable, or if it is not continuous everywhere, a number of difficulties will appear. The increase in the complexity of the solutions will produce the impossibility of explicitly obtaining these solutions. So we will study their properties through the study of their singularities.

**Fixed points** The more simple singularities, are the fixed points or double points which are the roots of the equation,

$$x - f(x) = 0\,. \tag{8.23}$$

For a dynamic system, directly described by an iterative process, a fixed point corresponds to an equilibrium state. In other cases, it may correspond to a periodic regime. An essential difference with the linear case thus appears: the iterative process 8.22 may have some fixed points at finite distances. Let $x = \alpha$ be such a point (the solution of 8.23 when taking $x = \alpha + X$,) where we put the origin at the fixed point $\alpha$. Let us suppose that $f(x)$ is differentiable, once at least, at $x = \alpha$. In the new iterative process obtained, we may separate the linear part in $X_n$ from the other terms $F(X_n)$, so that

$$X_{n+1} = SX_n + F(X_n), \quad F(0) = 0, \quad \lim_{X \to 0} F(X)/ \mid X \mid = 0\,, \qquad (8.24)$$

where $S = (df/dX)_{X=\alpha}$.

If we consider a sufficiently small domain $\mathcal{D}_\alpha$ surrounding the point $\alpha$, the linear approximation of 8.24 is

$$X_{n+1} = [df/dX]_{X=\alpha} \cdot X_n\,, \qquad (8.25)$$

when we write the first two terms of the Taylor development of $f(\alpha + X)$ and consider that $\alpha = f(\alpha)$. $S = [df/dX]_{X=\alpha}$ is the multiplicator of the fixed point $X = \alpha$.
As in the linear case, $S$ characterizes the behaviour of the series generated by 8.22 as $n \to +\infty$, but here, only for initial positions inside $\mathcal{D}_\alpha$.

**Example** Let us consider the iterative process equation,

$$x_{n+1} = x_n + ax_n^i, \quad i = 2, \text{ or } 3\,. \qquad (8.26)$$

For $i = 2$, the difference $x_{n+1} - x_n$ shows that the fixed point O $(x = 0)$ is an attractor of a series generated from $x_0 < 0$, and is a **repeller** for $x_0 > 0$. The fixed point O $(x = 0)$ is thus instable.
For $i = 3$, when $a > 0$, O is a repeller for a series generated from $x_0 < 0$, or $x_0 > 0$. When $a < 0$, O is an attractor for any $x_0$ taken in the neighbourhood. O is a repeller or instable in the first case, an attractor or asymptotically stable in the second case.
In this example, the stability of the fixed point O is determined by the the nonlinear term $ax^i$.

**Cycles** The fixed points are not yet the only possible singularities for a nonlinear iterative process.
Let us consider the iterative process which is $k$ iterated times,

$$x_{n+k} = f_k(x_n) = f_{k-1}[f(x)] = f[f_{k-1}(x)], \quad k = \text{integer} > 1\,, \qquad (8.27)$$

and which represents the transformation $T_k$ as constructed from equation 8.22.
The fixed points of 8.27, which are not fixed points of 8.22, are singularities of the last iterative process. They are solutions of the equation,

$$x - f_k(x) = 0\,, \tag{8.28}$$

and they are called **cycles**.

More precisely, a point $\alpha$ is called a point of an **order k cycle** of the transformation $T$, if it is a fixed point of $T^k$, without being a fixed point of $T^l$, $1 \le l < k$ ($l$ and $k$ integers), so that

$$T^k\alpha = \alpha, \quad T^l\alpha \neq \alpha, \quad 1 \le l < k\,. \tag{8.29}$$

Points $\alpha_2 = T\alpha$, $\alpha_3 = T^2\alpha$, $\ldots$, $\alpha_k = T^{k-1}\alpha$ are points of cycle $k$, which is constituted of the $k$ points $\alpha_i$, $i = 1, 2, \ldots, k$, $(\alpha_1 = \alpha)$. A fixed point is, therefore, a $k = 1$ order cycle.
To a cycle, there is associated a multiplicator $S$, which is the same for all points of the cycle. Let $f(x)$ be differentiable at the points $\alpha_i$. Then

$$S = (df_k/dx)_{x=\alpha} = (df/dx)_{\alpha_1} \cdot (df/dx)_{\alpha_2} \ldots (df/dx)_{\alpha_k}\,, \tag{8.30}$$

is the multiplicator of the cycle and is the same for all points $\alpha_i$.
For a dynamic system, a cycle corresponds to a periodic regime which is a subharmonic of a reference periodic regime relative to a fixed point.

**Non uniqueness of $T^{-1}$, endomorphism** The property of uniformity of the inverse transformation $T^{-1}$ which is always satisfied in the linear case, is now a property which only exists for some forms of the uniform function $f(x)$ of 8.22.
Any point $x_n$ may have many rank one antecedents depending on the domain where it is. When $T^{-1}$ gives different determinations, we say that $T$ is an **endomorphism**.

An example is given by the iterative process,

$$x_{n+1} = Sx_n + x_n^2\,, \tag{8.31}$$

which may be written,

$$x_n = 0.5[-S \pm \sqrt{S^2 + 4x_{n+1}}] \tag{8.32}$$

and thus gives two determinations. For initial points $x_0$, $x_0 > -S^2/4$, there are two antecedents $x_1^{(1)}$, $x_1^{(2)}$, given by 8.32. For $x_0 < -S^2/4$, points $x_0$ have no real antecedents. The point $x_0 = -S^2/4$ has two identical first rank antecedents. Such a point is called a **critical point**.

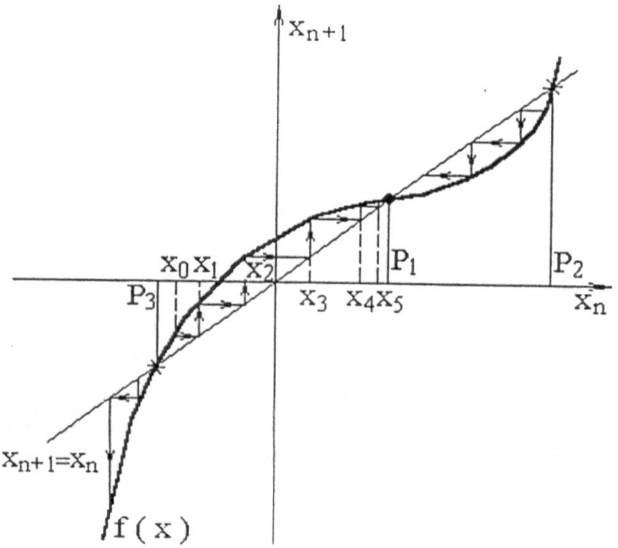


Figure 8.5: *Attractive domain of an attractive fixed point.* Using the Koenigs-Lemeray construction we determine the properties of the boundaries $\mathcal{F}$ of the domain $\mathcal{D}$. $P_1$ is attractive, $P_2$ and $P_3$ are repulsive. The domain $\mathcal{D}$ is constituted by the segment $P_2P_3$ (after Mira, 1987).

**Domain of stability of an attractive fixed point**

Such a domain $\mathcal{D}$ includes the set of points $x_0$ from which, equation 8.22 generates an iterative series which asymptotically converges to the attractive fixed point $\alpha$. In the linear case $\mathcal{D}$ is the entire $x$ axis, excluding the infinite point. In the nonlinear case, a more complicated situation may arise because we may have more than one singularity at a finite distance. For a dynamic system which can be described by an iterative process, $\mathcal{D}$ characterizes the domain of perturbations associated with shifts in initial conditions but with no modification in the qualitative behaviour of the system. To better understand the properties of the boundaries $\mathcal{F}$ of the domain $\mathcal{D}$ let us present the **Koenigs-Lemeray construction** which is illustrated in figure 8.5.

In the plane $x_n, x_{n+1}$ we consider any given function corresponding to the iterative process $f(x_n)$ as it is given by the equation 8.22. The bissectrix of the first quadrant is drawn. Then starting from an initial point $x_0$ we construct $x_1$, the value of $f(x_0)$. To do this, we use point $A$ on the curve, the abscissa of which is $x_0$ and then, observing that the ordinate of point $A$ is $x_1$, we obtain point $B$ on the bissectrix, the abscissa of which is $x_1$. This process is continued for all the following points.

Note that the fixed points are intersection of $f(x)$ with the bissectrix of the first quadrant. In figure 8.5 we notice that $P_1$ is attractive and that $P_2$ and $P_3$ are repulsive. So the domain $\mathcal{D}$ is constituted by the segment $P_2P_3$ which contains $P_1$.

In figure 8.6, we have an example where the stability domain, or attractive domain, is on the segment $P_2P_{2,-1}$ where $P_{2,-1}$ is the antecedent of $P_2$. Here, the attractive fixed point is $P_1$.

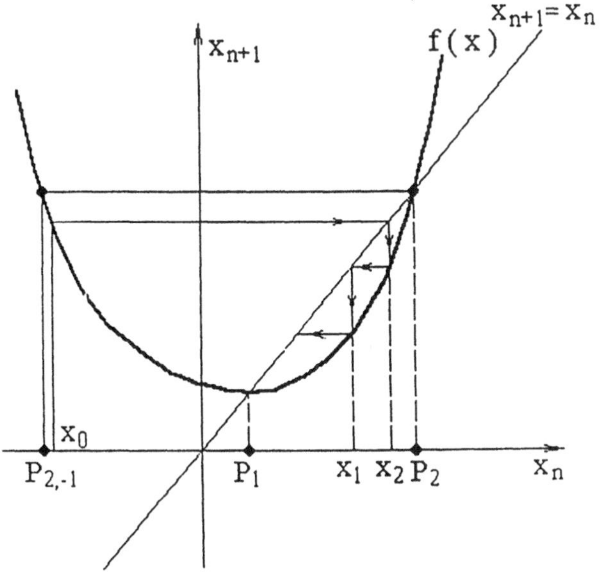


Figure 8.6: *Attractive domain of a fixed point.* In this example the boundaries $\mathcal{F}$ of the domain of stability $\mathcal{D}$ (or attractor) is constituted by the repulsive fixed point $P_2$ and its antecedent $P_{2,-1}$ which differs from $P_2$ (after Mira, 1987).

In the case of endomorphism, the situation becomes more complex. Cycles of order greater than 2 may be present and the number of singularities may be infinite and produce complex dynamic behaviour. In this situation, attractive fixed points may coexist with an infinity of expulsive cycles. Such segments are called **stochastic segments** or **chaotic segments**. The boundary $\mathcal{F}$ between attractive domains of two attractive fixed points may have a very complex structure, as shown on the figure 8.7. We can see on the $x_n$ axis attractive domain segments. There is an infinite series of possible antecedents, $Q^1_{-2}Q^2_{-1}$, $Q^1_{-3}Q^2_{-2}$, etc, the lengths of which $\rightarrow 0$ when $n \rightarrow \infty$.

There is an incertainty in the evolution of the series generated by $T$ from an initial condition $x_0$ towards one or another of the attractors.

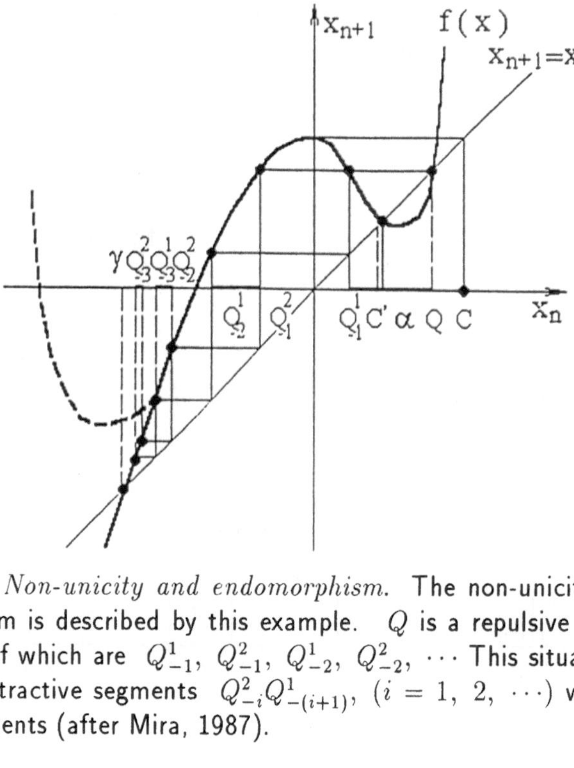


Figure 8.7: *Non-unicity and endomorphism.* The non-unicity of $T^{-1}$ and endomorphism is described by this example. $Q$ is a repulsive fixed point the antecedent of which are $Q^1_{-1}$, $Q^2_{-1}$, $Q^1_{-2}$, $Q^2_{-2}$, $\cdots$ This situation defines an infinity of attractive segments $Q^2_{-i}Q^1_{-(i+1)}$, $(i = 1, 2, \cdots)$ which are called chaotic segments (after Mira, 1987).

### 8.4.2 Bifurcation

We say that a solution of an iterative process corresponds to a **bifurcation** when there is a separation between two different qualitative behaviours of solutions produced by the effect of variations of a parameter or by a modification in the structure of the equation. More precisely, let us write the iterative equation,

$$x_{n+1} = f(x_n, \lambda), \tag{8.33}$$

where $\lambda$ is a real parameter, $f(x, \lambda)$ is a continuous function of both the real variable $x$ and of $\lambda$. If there exists $\epsilon > 0$ such that the structure of singularities of equation 8.33 is different for $\lambda < \lambda_0$, $\lambda > \lambda_0$ and $\lambda = \lambda_0$ ($|\lambda - \lambda_0| < \epsilon$), then the value $\lambda_0$ of the parameter $\lambda$ is called a **bifurcation value**.

A very classical example will summarize the above development of the concepts of iterative processes, first return maps, Poincaré sections and bifurcations. It concerns the **Feigenbaum attractor** or quadratic mapping.

### The example of the quadratic mapping or logistic map, bifurcations, sub-harmonic cascade and chaos

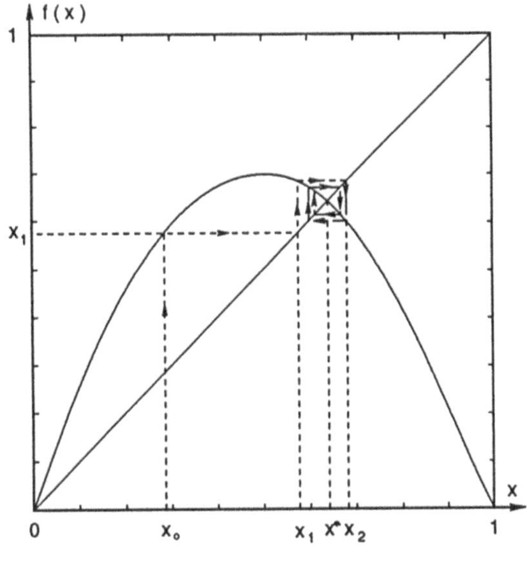


Figure 8.8: *The quadratic mapping for* $\mu < 3/4$. The point of intersection of the parabol and the first bissectrix the absciss of which is $x^*$ is an attractor towards which all iterates converge (after Bergé *et al.*, 1984).

We start from the continuous equation, [57, 58]

$$f(x) = 4\mu x(1 - x), \quad x \in [0,\ 1], \quad 0 < \mu \leq 1\,. \tag{8.34}$$

To define a first return map, we consider the iterative relation

$$x_{n+1} = 4\mu x_n(1 - x_n) = f(x_n) \tag{8.35}$$

which gives, for any $x_k$, a point $x_{k+1}$ on $[\,0,\ 1\,]$.
Let us draw the graph $f$ for a given value of $\mu$. This curve is a parabola crossing points $x = 0$ and $x = 1$, its maximum is $\mu$ when $x = 0.5$. Lets draw the first bissectix which intersects the parabola in two points, the abscissa of which are the roots of the equation $x = 4\mu x(1-x)$ and are given by $x = 0$ and $x = x^\star = (4\mu - 1)/4\mu$ (see figure 8.8).
This mapping which is apparently very simple, will lead to very complex situations as the $\mu$ value varies.

The tangent slope to the second intersection point is $f' = 2 - 4\mu$. Its value $f' = -1$ corresponds to $\mu_1 = 3/4$ which characterizes a very important change in the mapping.

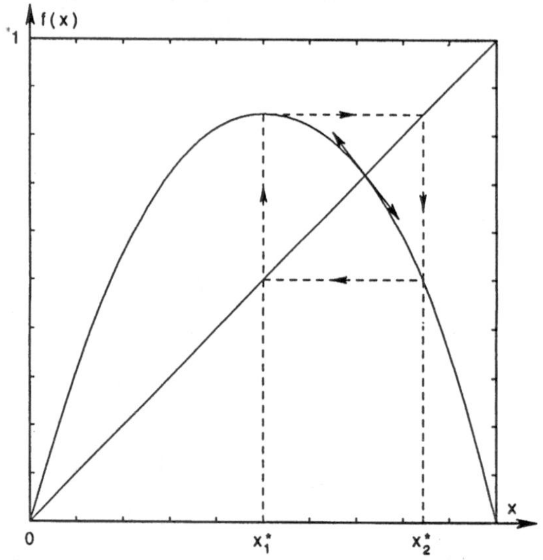


Figure 8.9: *Quadratic mapping for* $\mu = 0.8$. The attractive point is replaced by $x_1^\star$ and $x_2^\star$ such as $x_2^\star = f(x_1^\star)$ and $x_1^\star = f(x_2^\star)$ (after Bergé *et al.*, 1984).

1. When $\mu < 3/4$, an application of the Koenig-Lemeray construction allows us to see that an iterative process starting from any $x_0 \in [0,\ 1]$ leads to the point $x^\star$ which is the absciss of the intersection point of the parabola $f(x)$ with the first bissectix (figure 8.8).

   We will say that this point is a stable fixed point or an **attractor**, and we will also say that it is a **period one attractor** (by reference, in the phase space, to a system the limit cycle of which has a Poincaré section given by this point).

When $\mu < 3/4$ the absolute value of the tangent slope is less than one.

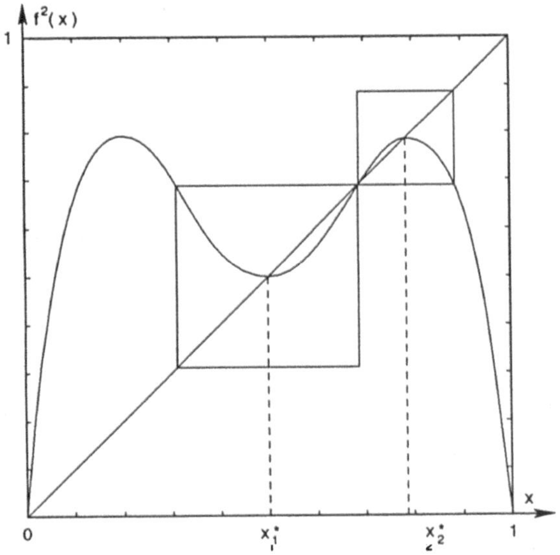


Figure 8.10: *Graph of the mapping* $g(x) = f(f(x)) = f^2(x)$ *for* $\mu = 0.8$. The points $x_1^\star$ and $x_2^\star$ are attractors. Iterative process pass alternatively to one to the other. In each square the structure is similar to the one of the previous figure (after Bergé *et al.*, 1984).

2. $\mu = 3/4$ or $\mu = 3/4 + \epsilon$. The absolute value of the tangent slope is one or more than one. The figure 8.9 shows that the point $x^\star$ is becoming unstable when $\mu$ is crossing the $3/4$ value.

   For example, on figure 8.9, where $\mu = 0.8$ we observe that the point $x^\star$ is replaced by two points $x_1^\star$ and $x_2^\star$ such that

   $x_2^\star = f(x_1^\star)$ and $x_1^\star = f(x_2^\star)$, that is to say

   $x_2^\star = f(f(x_2^\star))$ and $x_1^\star = f(f(x_1^\star))$.

   This means that there are two fixed points in the graph $f^2(x)$, defined by the intersections of the first bissectrix and the curve $f^2(x)$, and the mapping concerns, alternatively, both of them (see figure 8.10).

   They constitute a **period 2 attractor**; it is necessary to have two periods on the limit cycle to come back again on the same point of the Poincaré section.

To summarize, when $\mu$ crosses the $3/4$ value, a period one attractor is replaced by a period two attractor. We say that there is a **bifurcation.**

3. $\mu$ increases, the curves $f$ and $f^2$ change, and points $x_1^\star$ and $x_2^\star$ are becoming unstable when the absolute value of the tangent slope to the $f^2$ curve crosses over the value 1. If we enlarge $f^2$ around $x_1^\star$ and $x_2^\star$, we notice that the situation is compatible with the critical case $x^\star$ on $f$.

   To each of the two points will be associated two new points, that is to say four points on $f^4$. The critical value of $\mu$ corresponding to this new change is $\mu_2 = (1+\sqrt{6})/4 = 0.86237\ldots$ (figure 8.11).

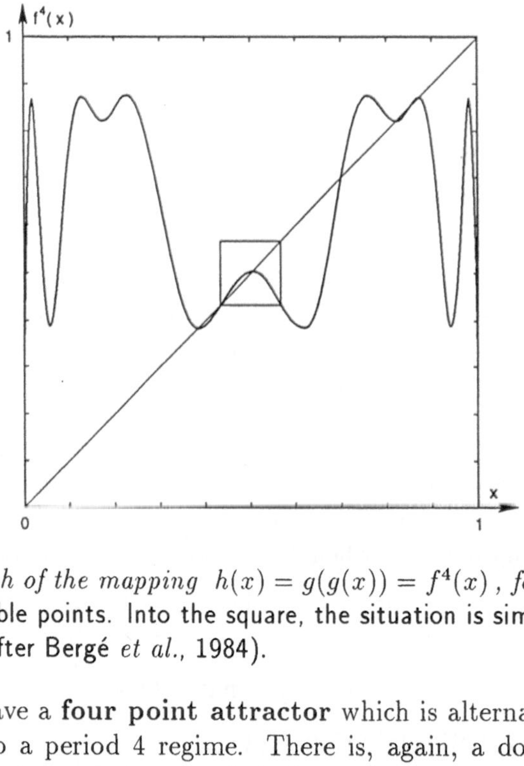


Figure 8.11: *Graph of the mapping* $h(x) = g(g(x)) = f^4(x)$, *for* $\mu = 0.875$. There are four stable points. Into the square, the situation is similar to that in previous figures (after Bergé *et al.*, 1984).

Thus, we have a **four point attractor** which is alternatively visited according to a period 4 regime. There is, again, a doubling of the period through a **sub-harmonic bifurcation.**

This operation may be repeated an infinite number of times as $\mu$ values continue to increase giving a **bifurcations cascade** (see figure 8.12).

There is, in this development, an analogy between the structures at different scales, and a Cantor set. When $\mu$ increases, a set of attractors

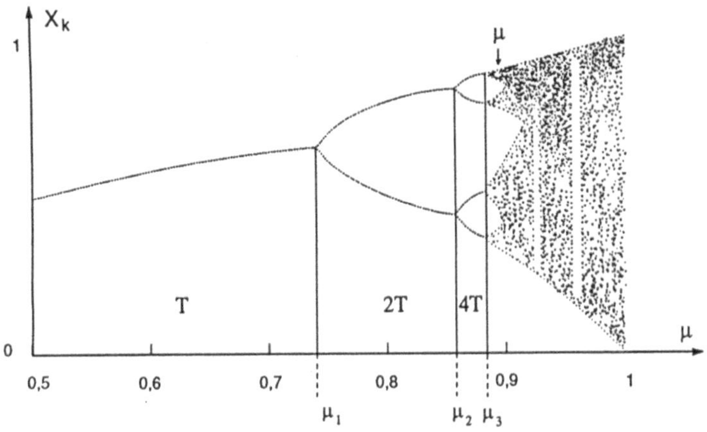


Figure 8.12: *Bifurcations cascade for a quadratic equation when $\mu$ increases.* At each bifurcation, the number of fixed points (attractor) is doubling (after Feigenbaum, 1978 ; Bergé *et al.*, 1984).

is created, the period of which increases like $2^l$ where$l$ is an integer between 0 (for $\mu \leq 0.3/4$) and infinity.

It has been shown numericaly [57, 58] (Feigenbaum, 1978, 1979) that $\mu_\infty = 0.892486418\ldots$ and that the convergence to $\mu_\infty$ depends on a scaling law,

$$\lim_{i\to\infty} \frac{\mu_i - \mu_{i-1}}{\mu_{i+1} - \mu_i} = \delta\,. \tag{8.36}$$

The scale reduction factor $\delta$ is a scale invariant factor which does not depend on $f$ and has the value $\delta = 4.6692016\ldots$ (Feigenbaum, 1978, [57]). $\delta$ is called the **Feigenbaum constant**.

When $\mu > \mu_\infty = 0.892\ldots$, we enter a zone where periodic and aperiodic attractors are present. When aperiodic, we are in a chaotic behaviour domain where iterates of $f$ give succession of values of $x$ which are never the same and which depend on the initial condition $x_0$. This property of Initial Conditions Sensibility is characteristic of strange attractors which will be studied in more details in the next section.

### Dynamic Bifurcations

Let us complete this study in its theoretical aspect. For that, we follow Beasens (1991) [19] who studied dynamic bifurcations in the case of period-doubling trees. The problem is one of dealing with dynamic systems that are one-parameter families of two-dimensional mappings

$$F_v(x_n, \lambda_n) : \begin{cases} x_{n+1} &= f(x_n, \lambda_n) \\ \lambda_{n+1} &= \Lambda_v(\lambda_n), \quad 0 \mid v \mid \ll 1, \end{cases} \tag{8.37}$$

where $v$ is the sweep and gives the non autonomous logistic map developped by Kapral and Mandel [113] in the following equation

$$\begin{aligned} x_{n+1} &= \lambda_n x_n (1 - x_n) \\ \lambda_{n+1} &= \lambda_n + v, \quad 0 <\mid v \mid \ll 1; \end{aligned} \tag{8.38}$$

In both forward ($v > 0$) and backward ($v < 0$) sweep, the orbit shows a transition from period $2^n (resp. 2^{n+1})$ to period $2^{n+1} (resp. 2^n)$ which is delayed, and only a finite (small) number of doublings are observed (see [19]).

## 8.5 Attraction and Repulsion Cycles

In the previous section we introduced the concept of an attractor in a discrete dynamic system.
Let us now developp this concept for a mechanical example.
Let us consider an oscillating pendulum, such as a pendulum clock, which is maintained by any power system (electro-magnet impulsions, mechanical spring, etc) that counters the frictional energy loss of the system. A steady state is obtained when the clock is working. In the two-dimensional phase space we may define a **limit cycle** $C$ (figure 8.13). Any point on $C$, the coordinates of which are $\theta(t)$, $d\theta/dt(t)$, defines completely the system state at the instant $t$.
If we keep away from (outside or inside) the limit cycle area, the system will automatically restore its initial position. When the figurative point is outside the limit cycle, the friction energy loss is higher than the power energy and the amplitude of oscillations decreases until the limit cycle is attained. On the contrary, when the figurative point is inside the limit cycle, the frictional

energy loss is lower than the power energy and the amplitude of oscillations increases until the limit cycle is again attained.
This experiment gives a simple image of the attraction phenomenom. The limit cycle in this example may be considered as an attractor in the phase space of the dynamic system.

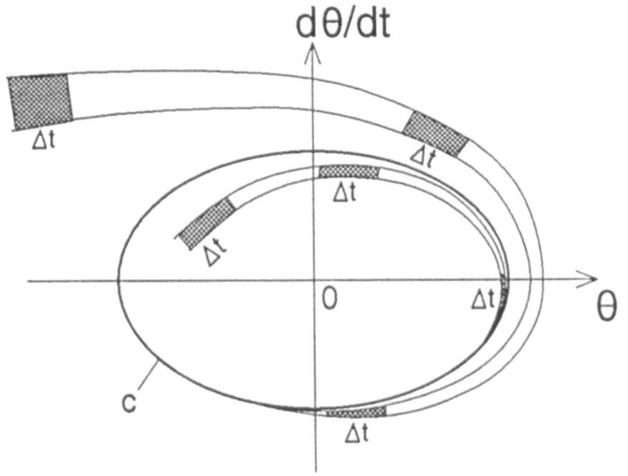


Figure 8.13: *Areas contraction.* Limit cycle in the phase space $\theta(t)$, $d\theta/dt(t)$ and the contraction of areas phenomenom. Each area corresponds to the same value of time $\Delta t$. The decrease of the area implies loss of information.

### 8.5.1 Areas contraction and consequences

In figure 8.13, we consider two adjacent trajectories starting from two different (but nearby) initial positions that are outside the limit cycle area. During the nth interval of time of duration $\Delta t$, we define the area $A_n$ between these two trajectories. In figure 8.13 we see that $A_1$ is larger than $A_2$ which is larger than $A_3$ (all having the same value of $\Delta t$). The areas are getting smaller and smaller with time as the phase trajectories converge on the limit cycle. We say that there is an area contraction. The area approaches zero when the trajectory is on the attractor.
A first consequence of area contraction is the loss of information which was contained in the area. When the attractor is attained, the information is completely erased. The exact position of the initial point has been forgotten.
A second important porperty of attractors is the following. To define the phase trajectory in the phase space, we need two coordinates, $\theta$ and $d\theta/dt$.

When the trajectory has gone asymptotically to the limit cycle, only one trajectory remains and it is possible to consider only one curvilinear coordinate along the limit cycle curve to completely determine the system state. In this simple example, we notice a general property of attractors. The dimension $d$ of the attractor is lower than the phase space dimension $n$, that is to say to the number of degrees of freedom of the dynamic system, $d < n$.

## 8.6 Fractal Attractors, Basin of Attraction, Repellers

A mechanical system, like a central field oscillator, presents a Fourier spectrum of the variations of one of its coordinate with a frequency continuum, like a noise spectrum [37]. This characterizes the erratic motion behaviour of the system. One may estimate the disorder rate, by introducing a function which measures the likeness of $X$ at time $t$ given its value at $t + \tau$. An example is the auto-correlation function

$$C(\tau) = \frac{1}{t_2 - t_1} \int_{t_1}^{t_2} X(t) \cdot X(t + \tau) dt \,. \tag{8.39}$$

or

$$C(\tau) = \langle X(t) \cdot X(t + \tau) \rangle \,. \tag{8.40}$$

We may construct the function $C(\tau)$ by varying $\tau$. It has been shown (Wiener, Kintchin) that $C(\tau)$ is the Fourier transform of the power spectrum. If $X(t)$ is constant, periodic or quasi-periodic, $C(\tau)$ will remain different from zero as $\tau$ tends to infinity. In this case, the system behaviour is predictible.

On the contrary, when the regime is chaotic aperiodic, the power spectrum having a continuous part, $C(\tau)$ tends to zero as $\tau$ increases. This means that temporal similarity of the signal with itself decreases and disappears for significantly long intervals of time. It results from this, that even knowing $X(t)$, on a very long interval of time, doesn't preclude predicting future behaviour of $X(t)$. One says that the system has an impredictible chaotic regime by loss of internal similarity.

We shall see that the degree of impredictibility, in fact, is less absolute than this implies. It has been shown [115] (Keilis-Borok, 1990) that on the Lorenz attractor, the jump of a representative point on the phase trajectory on one lobe of the attractor to the other does occur with a certain repetitivity in a small domain of the same lobe. Recent works [171] [170] have shown that evolution of orbit of chaotic attractor may be predictible on short time intervals. We shall study in detail this point in the section of control of chaos (10.4).
But nevertheless, the main consequence of the chaotic behaviour is that two trajectories, initially close, diverge very rapidly on the attractor, giving a multitude of final positions and they lose any similarity after only a short interval of time (figure 8.14).

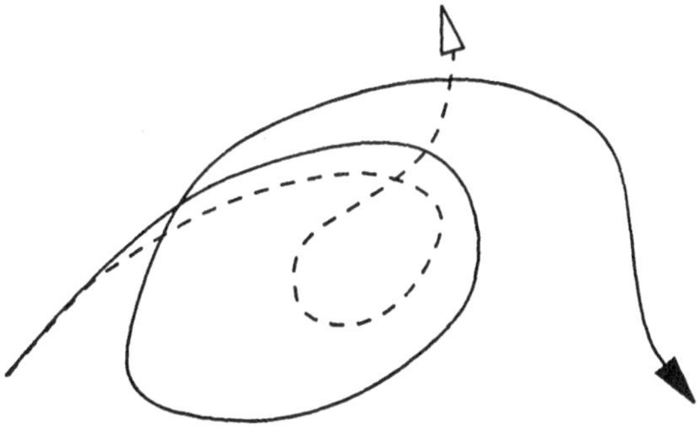

Figure 8.14: *Initial conditions sensibility.* Divergence of two neighbour phase trajectories which gives non predictible final position (after Dubois, 1995).

The amplification of divergence between trajectories on the attractor is called Initial Conditions Sensibility, I.C.S..
There is a sort of counterpoint between the attraction phenomenom which erases informations (area contraction) and the divergence on the attractor which means creation of informations.

A third point characterizes this special class of chaotic attractor. To have ICS in a dynamic system, it is necessary that the dimension of the attractor $d > 2$. But, in a dissipative system with area or volume contraction, the volume of the attractor in the euclidian phase space has to be zero. This

means that in a tridimensional regime its dimension is $d < 3$. So an attractor which may represent a chaotic regime with ICS has a dimension $2 < d < 3$. This is impossible in a euclidian space, but is possible in a non-integer or fractal dimension space.
To summarize, dissipative dynamic systems may be chaotic for a phase-space dimension equal to or greater than 3. The chaos with a small number of degrees of freedom is due to ICS of the trajectories which cover the attractor. These attractors which are called chaotic or strange attractors [153] have three main properties,

1. The phase trajectories (or orbits) are attracted by a geometrical object called an attractor.
2. Pairs of neighbour trajectories diverge on the attractor (ICS).
3. The dimension of the attractor is fractal.

**Remark**

More precise mathematical definitions of attractors and repellers were given in sections 8.2 and 8.3.

## 8.7 Lyapunov Exponents

In order to introduce Lyapunov exponents, let us first define the **Jacobian** of a transformation.
We start from a **linear transformation** which maps a point $M\ (x,y)$ in $Ox, Oy$ to a point $P\ (z,t)$ in $Oz, Ot$. The transformation is

$$z = \alpha x + \beta y,$$

$$t = \gamma x + \delta y,$$

where $\alpha$, $\beta$, $\gamma$, $\delta$ are real numbers.
We consider the square $OM_1MM_2$ on $Ox, Oy$ $M_1(1,0)$, $M_2(0,1)$, $M(1,1)$ (figure 8.15).
The transformation of the square is a parallelogram in $Oz, Ot$ $OP_1PP_2$, $P_1(\alpha,\gamma)$, $P_2(\beta,\delta)$, $P(\alpha+\beta,\ \gamma+\delta)$.

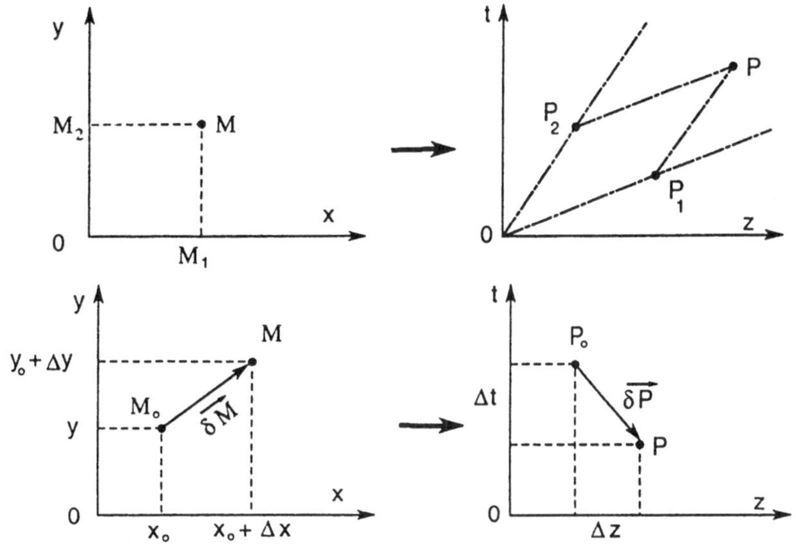


Figure 8.15: *Linear transformation of a point and vector.* We transform $M(x,y)$ to a point $P(z,t)$, then the vector $\overrightarrow{\delta M}$ to $\overrightarrow{\delta P}$ (after Bergé *et al.*, 1984).

The area of the parallelogram is the absolute value of the determinant

$$\begin{pmatrix} \alpha & \beta \\ \gamma & \delta \end{pmatrix} = \alpha\delta - \beta\gamma\,. \tag{8.41}$$

When $\mid \alpha\delta - \beta\gamma \mid > 1$, there is a dilation of areas,
When $\mid \alpha\delta - \beta\gamma \mid < 1$, there is contraction of areas.
Now let us consider a **non-linear transformation** $z = f(x,y)$; $t = g(x,y)$. In the neighbourhood of a point $M_0$ $(x_0, y_0)$ for small $\Delta x$, $\Delta y$ we delete second order terms, so that,

$$df = \frac{\partial f}{\partial x}dx + \frac{\partial f}{\partial y}dy = f'_x dx + f'_y dy \tag{8.42}$$

and

$$\Delta z \approx f'_x \Delta x + f'_y \Delta y\,.$$

with a similar relation for $\Delta t$.
Thus, the vector $\overrightarrow{\delta P}$ transformed from the vector $\overrightarrow{\delta M}$ in the neighbourhood of $M_0$ is

$$\overrightarrow{\delta P} \approx \begin{pmatrix} f'_x(x_0,y_0) & f'_y(x_0,y_0) \\ g'_x(x_0,y_0) & g'_y(x_0,y_0) \end{pmatrix} \overrightarrow{\delta M} = \mathbf{J}\overrightarrow{\delta M}\,. \tag{8.43}$$

Matrix **J** is called the **Jacobian** of the transformation at $x_0, y_0$.
When $|\mathbf{J}| > 1$ there is dilation of areas around $M_0$.
When $|\mathbf{J}| < 1$ there is contraction of areas.
Now let us follow Farmer *et al.* [55], and consider a mapping in a $p$ dimensional space, $x_{n+1} = F(x_n)$ where $x$ is a $p$ dimension vector.
To define the **Lyapunov numbers**, let $J_n = [J(x_n), J(x_{n-1}), \ldots, J(x_1)]$ where $J(x)$ are the Jacobians of the mapping, $J(x) = \partial F / \partial x$, and $j_1(n) \geq j_2(n) \geq \ldots \geq j_p(n)$ be the eigenvalues of $J_n$.
The Lyapunov numbers are

$$\lambda_i = \lim_{n \to \infty} [j_i(n)]^{1/n}, \quad i = 1, 2, \ldots, p. \tag{8.44}$$

where the only positive $n^{th}$root is taken.
This definition was proposed by Osedelets [139]. Conventionaly, we define the ordering $\lambda_1 \geq \lambda_2 \geq \ldots \geq \lambda_n$.
In the case of a $2D$ mapping, $\lambda_1$ and $\lambda_2$ are the axis of an elliptical transformation of an initial small circle. When the attractor is chaotic, neighbouring points diverge exponentially and, thus, one of the Lyapunov numbers is larger than the other. They represent the semi-axis of an ellipse (figure 8.16). This is an example of ICS.

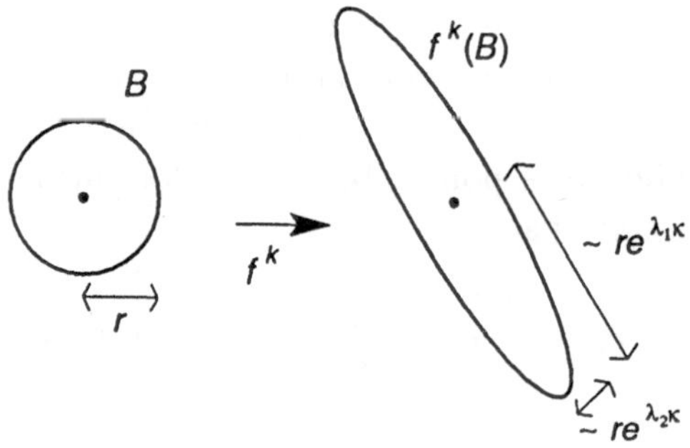


Figure 8.16: *Illustration of Lyapunov exponents in the phase space.* The transformation of an initial circle after $k$ iterates of a mapping $f$ (after Falconer, 1990).

The term **Lyapunov exponents** which are logarithms of Lyapunov numbers is also used.

### 8.7.1 Case of discrete time sampled physical system

Let us examine the more usual case in the study of dynamic system as developed by Hongre *et al.* (1995) [106].
In realistic situations, the parameters of a physical system are observed at discrete times sampled every $\tau_s$. This leads to discrete flow in the phase space $\mathbf{R}^d \to \mathbf{R}^d$ labeled by an observation index $\mathbf{x}(n) = \mathbf{x}(t_0 + n\tau_s)$

$$\mathbf{x}(n+1) = \mathbf{F}(\mathbf{x}(n)). \tag{8.45}$$

Dependence of the dynamics on initial conditions is described by the spectrum of **Lyapunov exponents** of the vector field $\mathbf{F}(\mathbf{x})$.
Let us consider an infinitesimal $d$-sphere with principal axes $\mathbf{e}_i(0)$, $i = 1, \ldots, d$, each representing a deviation from the $\mathbf{x}(0)$ at time $t = t_0$, $(n = 0)$. As the sphere evolves under the action of non-uniform flow, it will become a $d$-ellipsoid with principal axis $\mathbf{e}_i(n)$ (infinitesimality allows to consider only linear deformations).
There exist $d$ different limits

$$\lambda_i = \lim_{n\to\infty} \frac{1}{n\tau_s} \ln \frac{\| \mathbf{e}_i(n) \|}{\| \mathbf{e}_i(0) \|} \, i = 1, \ldots, d, \tag{8.46}$$

which are called 1-dimensional Lyapunov exponents, usually ordered such that $\lambda_1 \geq \lambda_2 \geq \ldots \geq \lambda_d$ (we denote by $\| \cdot \|$ Euclidean norm of a vector). They are the average exponential rates of divergence or convergence of nearby orbits in phase space. A positive Lyapunov exponent reflects an exponential divergence of orbits in a certain direction, which is one of the characteristic features of **deterministic chaos**: the long-term behaviour of an initial condition that is specified with any uncertainty cannot be predicted (amplification of noise).

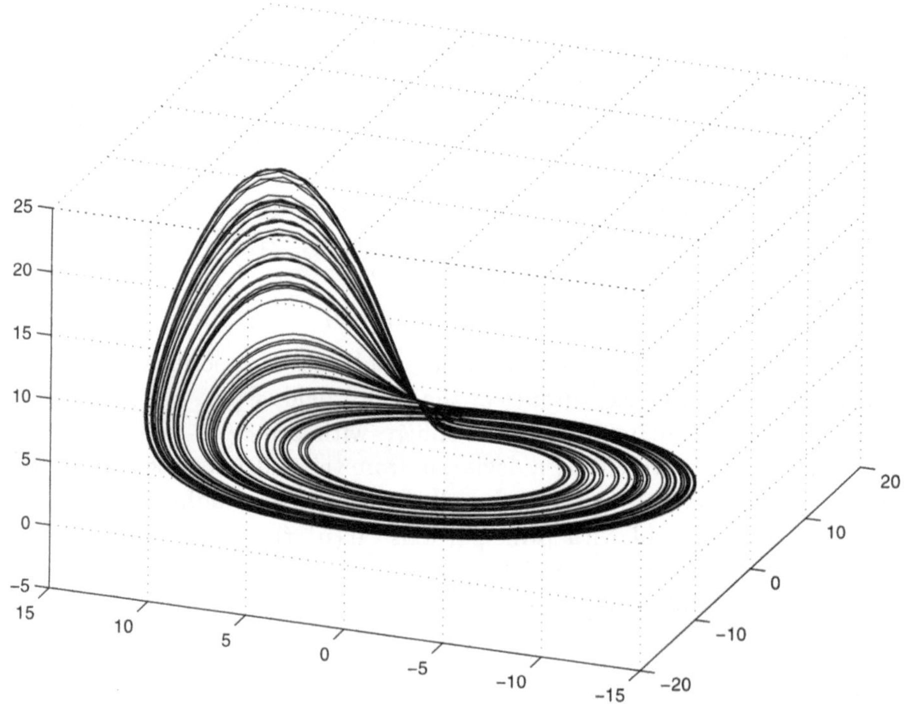


The Rössler Attractor – The parameters (see the text) are: $a = 0.385$ ; $b = 2$ ; $c = 4$. The perspective representation in the phase space shows 10,000 points discrete phase trajectories (after David AUBERT, 1997).

# Chapter 9

# Geometry of Attractors

## 9.1 Classical Examples of Strange Attractors

There have been many studies devoted to strange attractors [125, 101, 126, 151, 154, 127, 76, 20, 54]

We'll only give here some basic features of a few of them. At the present time, they are generally used to test any new study dealing with non-linear dynamics such as our previous investigations on volcanological data files [45, 46] (Dubois and Cheminée, 1988, 1991; Sornette *et al.*, 1991) or geomagnetic data files [50, 106] (Dubois and Pambrun, 1989; Hongre *et al.*, 1994).

All of these particular attractors are presented here because we want to systematically apply to them our tools for the study of natural dynamic systems; for example, the Lorenz attractor in Lyapunov exponents application on geomagnetic series processing, or, the Hénon attractor for the technique of controle of chaos etc.

### 9.1.1 The Lorenz attractor

The Lorenz model (1963) has, historically, a very big importance as it marks the reintroduction after the Poincaré ideas, of the non linear dynamics approach in modern physics.

This model permits, after some simplifications, to formulate the dynamic behaviour of a fluid during its convection, under three differential equations which define a three dimensional flow $X$, $Y$, $Z$.

The differential system is written

$$\begin{cases} \dfrac{dX}{dt} = P_r(Y - X), \\ \dfrac{dY}{dt} = -XZ + rX - Y, \\ \dfrac{dZ}{dt} = XY - bZ, \end{cases} \tag{9.1}$$

where $P_r$ is the Prandtl number of the fluid, $P_r = \nu / D_T$ where $\nu$ is the fluid kinematic viscosity, $D_T$ the thermic diffusivity, (for liquids and water, $P_r$ is between 5 and 10 depending on the value of $T$, while for silicon oils, $P_r > 100$), $4\pi^2/(\pi^2 + q^2)$ where $q$ is the pulsation, in the direction of fluid movement, the period of which is $2\pi/q$.
To simplify the problem, the coefficients are generally taken as $P_r = 10; b = 8/3$ and leaving $r$ as the control parameter.

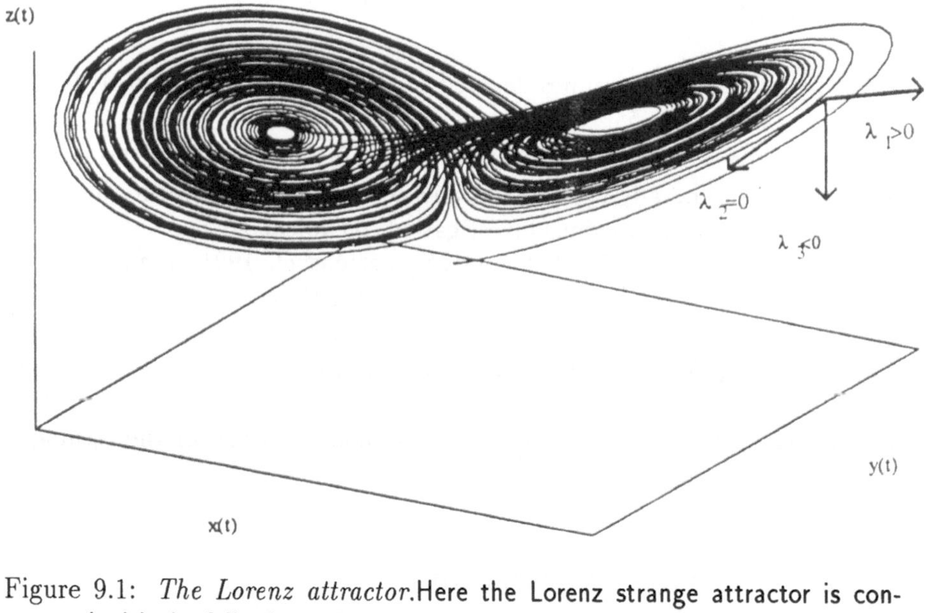


Figure 9.1: *The Lorenz attractor.*Here the Lorenz strange attractor is constructed with the following values of the parameter $Pr = 16$, $b = 4$, $r = 45.92$ (after Hongre *et al.*, 1994).

Generally, the system cannot be integrated. But it can be solved numerically when given initial values $X(0)$, $Y(0)$, $Z(0)$ by computing the flow $X(t)$, $Y(t)$, $Z(t)$, step by step on the trajectory. One may then observe that trajectories in the phase space remain always on the same geometrical object which is formed by two lobes (this object is very often reproduced in

the most current scientific reviews). It is a strange attractor, the Hausdorff dimension of which is 2.06, which is more than the euclidian dimension of a surface, so that it cannot be represented on a plane, but very easily in a $3D$ space where it doesn't fulfill a volume (figure 9.1).
By using the Lie derivative $1/V \cdot dV/dt$ where $V$ is the volume in the phase space defined by the phase trajectory, the volume variations under flow action are given by

$$\frac{1}{V}\frac{dV}{dt} = \sum_{i=1}^{n} \frac{\partial \dot{X}_i}{\partial X_i}, \tag{9.2}$$

where $X_i$ is the $i$th component of $\overrightarrow{X}$. Here

$$\frac{\partial \dot{X}_i}{\partial X} + \frac{\partial \dot{Y}_i}{\partial Y} + \frac{\partial \dot{Z}_i}{\partial Z} = -(P_r + b + 1) = \frac{41}{3}, \tag{9.3}$$

for $P_r = 10$ and $b = 8/3$.
After one time unit, the volume contraction is $e^{-41/3} \approx 10^{-6}$ which is very rapid. The model is highly **dissipative**
The Poincaré cross-section by the plane $Z = const., XOY$ permits first return mapping $Z_{k+1}$, $Z_k$ (see figure 9.1).

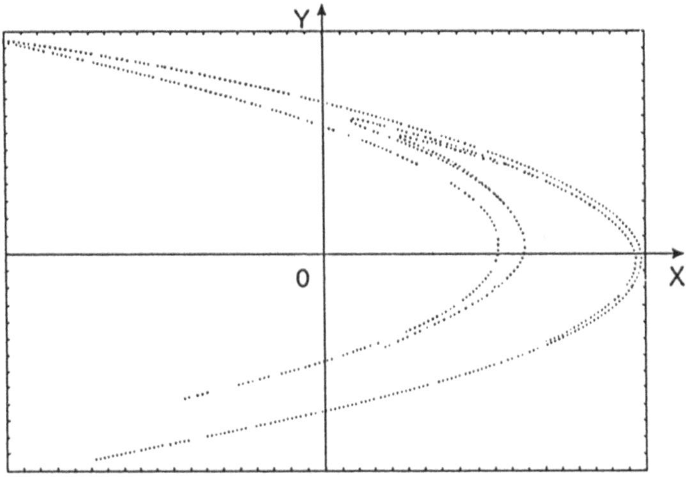


Figure 9.2: *The Hénon attractor.* The parameters are here: $\alpha = 1.4$; $\beta = 0.2$ (after Bergé *et al.*, 1984).

### 9.1.2 The Hénon attractor

By introducing simplifications to the thermal convection model, Hénon (1976), [101] proposed to replace the three differential equations system by a two-dimensional mapping. The points obtained in the plane may be considered as a Poincaré cross-section of a three-dimensional flow. The mapping works as following:

$$\begin{cases} X_{k+1} &= Y_k + 1 - \alpha X_k^2 \\ Y_{k+1} &= \beta X_k \end{cases} \tag{9.4}$$

where $\alpha$ and $\beta$ are constant which control linearity and dissipation. Generally we have $\alpha = 1.4$ and $\beta = 0.3$ (see figure 9.2)
The deterministic chaotic behaviour of this system is observed when starting from two neighbouring points close to the attractor, the initial distance between them being $\delta_0$. The distance $\delta$ between the $k$th iterated points (which is easy to compute) verifies a relationship $\delta = \delta_0 \exp(\lambda_1 k)$, where $\lambda_1$ is the first Lyapunov exponent, (here $\lambda_1 = 0.5$) which expresses the ICS property.
The property of area contraction in the phase space is also tested on the Jacobian value which is,

$$J = \begin{vmatrix} \dfrac{\delta X_{k+1}}{\delta X_k} & \dfrac{\delta X_{k+1}}{\delta Y_k} \\ \dfrac{\delta Y_{k+1}}{\delta X_k} & \dfrac{\delta Y_{k+1}}{\delta Y_k} \end{vmatrix} = \begin{vmatrix} -2\alpha X_k & 1 \\ \beta & 0 \end{vmatrix} = -\beta\,. \tag{9.5}$$

Because $\mid \beta \mid < 1$, the area contraction works.
The third property of strange attractors is self similarity and is also observed. The structure of the attractor is repeated when we successively reduce the scale of observation (figure 9.3).

### 9.1.3 Some other classical attractors

Among the more better known, one may mention:

#### The Rössler attractor

The Rössler attractor (1976) [151] has the geometrical aspect of a Mobius strip (figure 9.4). Its mathematical formalism is

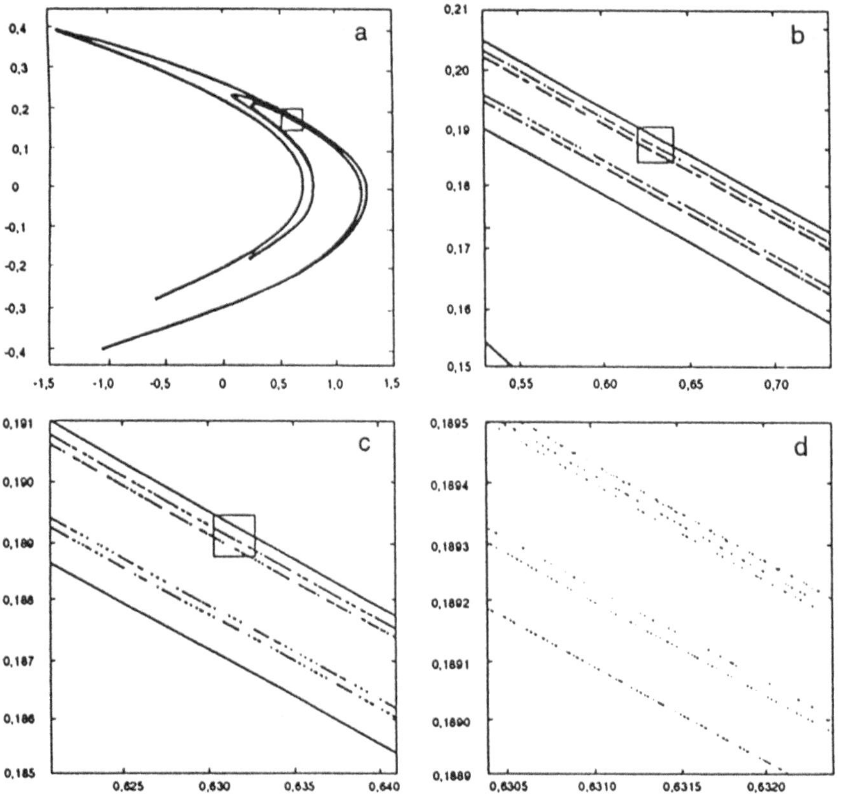


Figure 9.3: *Self similarity in the Hénon attractor.* The structure is repeated at different scales and illustrates self similarity of the attractor (after Hénon, 1976).

$$\begin{cases} \dfrac{dX}{dt} = -(Y+Z)\,, \\ \dfrac{dY}{dt} = X + aY\,, \\ \dfrac{dZ}{dt} = b + XZ - cZ\,. \end{cases} \tag{9.6}$$

When $a = b = 0.2$ and $c = 5.7$, the flow has a chaotic behaviour. When $b = 2$, $c = 4$ the structure of the attractor varies with $a$ from being a simple closed curve to becoming a double-loop curve and then a quadruple loop as $a$ increases. This recalls the behaviour of the logistic map with successive

doubling of the periods when the control parameter increases.
Rössler (1979) also defines a **Rössler-hyperchaos attractor**:

$$\begin{cases} \dfrac{dX}{dt} &= -(Y+Z)\,, \\ \dfrac{dY}{dt} &= X + aY + W\,, \\ \dfrac{dZ}{dt} &= b + XZ\,, \\ \dfrac{dW}{dt} &= cW - dZ\,, \end{cases} \tag{9.7}$$

for $a = 0.25;\ b = 3;\ c = 0.05;\ d = 0.5$, its Hausdorff dimension is 3.005.

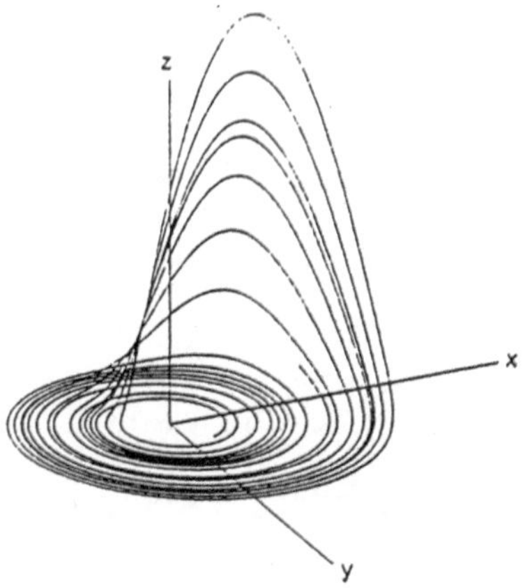


Figure 9.4: *The Rössler attractor.* This attractor looks like a Mobius ribbon in its phase space.

### The Feigenbaum attractor

This attractor [57] we have already studied in section 8.2 as the logistic mapping $f_\lambda(x) = \lambda x(1-x)$, or $f(x) = 4\mu x(1-x)$, $x \in [0,\ 1]$, $0 < \mu \leq 1$.

### The MacKey-Glass attractor

This attractor [126] characterizes dynamic systems which describe biological population behaviour:

$$\frac{dX}{dt} = \frac{aX(t+s)}{1 + [X(t+s)]^c} - bX(t)\,. \tag{9.8}$$

When, $a = 0.2$; $b = 0.1$; $c = 10$; and where $s$ is a given time delay, the dimension of the attractor is 6.7.

## 9.2 Correlation Function Method

In many dynamic systems we have only a series of values of one observable $X(t)$ and the problem consists of studying from these observations the attractor geometry of the dynamic system that generates $X(t)$. This is possible due to the properties of the attractor which were previously described. The essential question we address is to determine if any chaotic behaviour is resulting from a large number of degrees of freedom (i.e. random) or, if it is characterized by a small number which is the coin of deterministic behaviour of the system.
From the time evolution $X(t)$ it is possible to reconstruct an approximate phase space by the following standard procedure (also called the "method of lags" [140, 63, 173]) where one considers the set of $d_E$ new variables $\{X_j(t),\ j = 1 - d_E\}$ defined by

$$X_j(t) = X[t + (j-1)\tau] \quad j = 1 - d_E\,. \tag{9.9}$$

The phase space, of dimension $d_E$, the **embedding dimension,** is thus constructed with these different variables :

$$\begin{aligned} X_1(t) &= X(t),\ X_2(t) = X(t+\tau),\ X_3(t) \\ &= X(t+2\tau),\ \ldots,\ X_{d_E}(t) = X(t+(d_E-1)\tau)\,. \end{aligned} \tag{9.10}$$

In this scheme one considers the signal $X(t)$ to be independent of the same signal at a later time $X(t+\tau)$ where $\tau$ is an arbitrary constant called the delay. This is justified when the time correlation decays sufficiently rapidly on the time scale $\tau$. In this case the topology of the reconstructed phase space is the same as that of the true phase space. This does not mean that the attractor obtained in the new phase space is identical to that in the original phase space, but merely that the new representation of the attractor retains the same topological properties, which may often suffice for studying its essential characteristics. The mathematical justification of this scheme has been proven by Takens (1981) [173]. Intuitively, one may justify this method as follows [170]. When the attractor is simple, a fixed point or a limit cycle, the topological equivalence of the two representations is nearly obvious. It is no longer obvious for a more complicated attractor, and it is the last case that we will consider. The main idea is that the attractor of the system's dynamic evolution is by definition invariant under the dynamics. When one explores "inside" the attractor, the property of ergodicity holds. Since the reconstruction of a phase space by the method of time lags is in a sense the inverse of the Poincaré cross-section method, it is this property which ensures its validity. However, caution should be exercised, since precise bounds on the domain of validity for the equivalence of the two representations are still lacking. It is then useful to look at specific examples which teach how the method works.

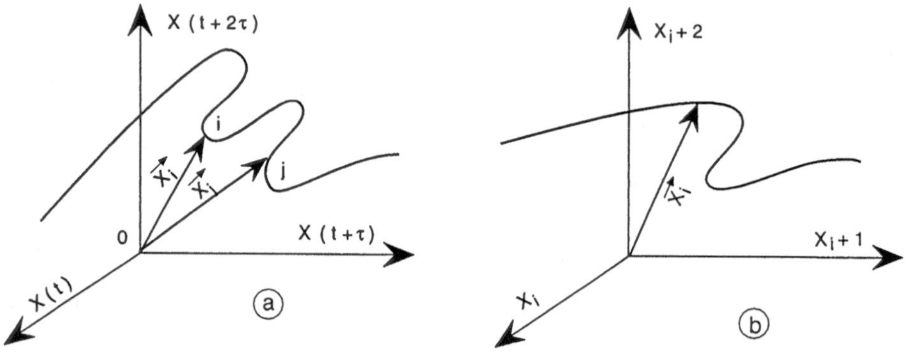


Figure 9.5: *Phase trajectory in a three-dimensional phase space.* Each point has coordinates $X(t)$, $X(t+\tau)$, $X(t+2\tau)$. In the second graph time is replaced by the order number (series of events, earthquakes or volcanic eruptions, $x_i$, $x_{i+1}x_{i+2}$ represent the time interval between two successive events). This pseudo phase space is equivalent to the previous one.

We may choose the time delay value $\tau$ as a multiple of the digitizing interval. We thus obtain a pseudo phase space, the simplest of them being constructed on the successive values of $X(t)$ according to the discretisation step interval, $X_i, X_{i+1}, \ldots, X_{i+p-1}$.
The figure 9.5 shows a 3-dimensional phase space. Each point of the phase space has for coordinates $X(t)$, $X(t+\tau)$, $X(t+2\tau)$. In the second graph, time is replaced by the order number. In this figure, we have the phase

trajectory as a series of successive discrete positions of the representative points.

We may assume that in a chaotic regime the positions of two points on the same trajectory which are separated by a long period of time are without any correlation between them. On the contrary, if all the points are situated on an attractor, there is a spatial correlation between them and it is possible to characterize this by a suitable correlation function.

Grassberger and Procaccia (1983) [75] have proposed to look at the asymptotic behaviour expressed as $r^v$ of the integrand correlation fonction

$$C(r) = \lim_{m\to\infty} \frac{1}{m^2} \sum_{i,j=1}^{\infty} H\,(r- \mid X_i - X_j \mid) \,, \tag{9.11}$$

where $H$ is the Heaviside function. This may be further written (Bergé *et al.*, 1984),

$$C(r) = \lim_{m\to\infty} \frac{1}{m^2} (\text{Number of pairs } i,\ j, \text{ the distance of which} \mid X_i - X_j \mid < r) \,, \tag{9.12}$$

where $i$ and $j$ are two indices of points on the trajectory.

If we verify that $\log C(r) = v \log r$, then $C(r) = r^v$.

We consider the phase trajectory for integer values of $d_E$, the phase dimension, 1, 2, 3, ..., $d_E$.

For each value of $d_E$, we compute $C(r)$ from the distances (figure 9.5)

$$\mid X_i - X_j \mid = \left[ (X_i - X_j)^2 + (X_{i+1} - X_{j+1})^2 + \ldots + (X_{i+(d_E-1)} - X_{j+(d_E-1)})^2 \right]^{\frac{1}{2}} \,. \tag{9.13}$$

which then gives the slope of the curve $\log C(r) = f(\log r)$.

In the case of a white noise, for example, or in the case of a dynamic system with a large number of degrees of freedom, the exponent $v$ increases continuously as $d_E$ increases. On the contrary, when $v$ becomes independent of $d$, the regime is deterministically chaotic and the attractor has the dimension of $v$. On a graph of $v$ against $d_E$, when we observe a constant value $v_s$ we say that we have saturation in the phase-space for the embedding dimension or phase-space dimension $d_s$ and for dimensions larger than $d_s$.

Atten and Caputo (1987) [8] have shown that a relation exists between $v_s$ and the saturation embedding dimension $d_s$, given by $d_s = 2v_s + 1$ (figure 9.6).

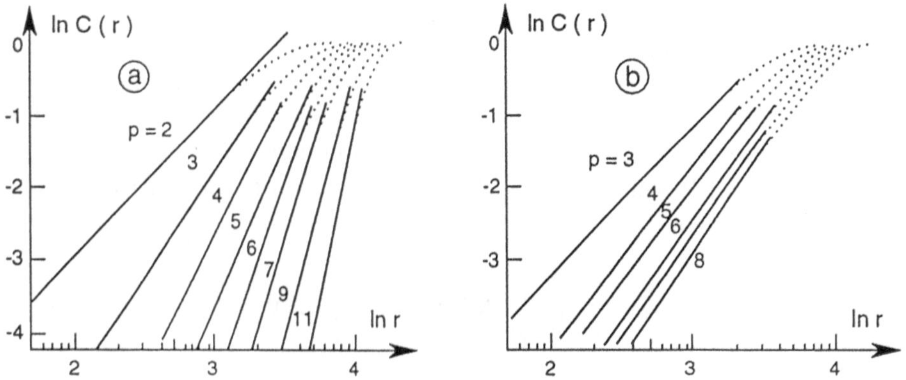


Figure 9.6: *Graph of saturation of $v$ against $d_E$*. When increasing $d_E$ the slope increases until it saturates in the case of a deterministically chaotic system.

**Remark**

The number of observable values $X(t)$ necessary to calculate the dimension of the attractor has to be large, several thousand or more. This minimum number depends of the attractor dimension. Ruelle (1990) [155] has shown that the number of values $N$ is linked to the attractor dimension by the relationships $N \gg 10^{D/2} \text{and} N \gg 42^D$.

## 9.3 What Time Delay?

Let us now examine a standard technique to reconstruct $d$-dimensional space from discrete scalar observations $X(t)$ after Ruelle (1987) [154], and the time-delay embedding

$$X(t) = [X(t),\ X(t+\tau),\ \ldots,\ X(t+(d_E-1)\tau]\ . \qquad (9.14)$$

The questions are: what time delay, $\tau$, and what embedding dimension, $d_E$, to use in this formula?

To begin, we will use the **mutual information criterion** [60, 106], to estimate the time delay $\tau$.

## 9.3.1 The mutual information criterion

The mutual information criterion measures the general statistical dependence between the time series $X(t)$ with a set of possible values $\mathcal{S} = \{s_i \mid \exists t\colon\ X(t) = s_i\}$ and the delayed time series $X(t+\tau)$ with a set of values $\mathcal{D} = \{d_i \mid \exists t\colon\ X(t+\tau) = d_i\}$.
The mutual information

$$\mathcal{I}(\mathcal{S},\mathcal{D}) = \mathcal{H}(\mathcal{S}) + \mathcal{H}(\mathcal{D}) - \mathcal{H}(\mathcal{S},\mathcal{D}) \tag{9.15}$$

is the sum of information from the individual time series measured in entropy bits,

$$\begin{aligned} \mathcal{H}(\mathcal{S}) &= -\textstyle\sum_i \mathcal{P}_{\mathcal{S}}(s_i)\log_2 \mathcal{P}_{\mathcal{S}}(s_i) \\ \mathcal{H}(\mathcal{D}) &= -\textstyle\sum_i \mathcal{P}_{\mathcal{D}}(d_i)\log_2 \mathcal{P}_{\mathcal{D}}(d_i) \\ \mathcal{H}(\mathcal{S},\mathcal{D}) &= -\textstyle\sum_i \mathcal{P}_{\mathcal{SD}}(s_i,d_i)\log_2 \mathcal{P}_{\mathcal{SD}}(s_i,d_i)\,. \end{aligned} \tag{9.16}$$

Here $\mathcal{P}_{\mathcal{S}}(s_i)$ is a probability that the series $X(t)$ take a value $s_i$, and $\mathcal{P}_{\mathcal{SD}}(s_i,d_i)$ is a probability that the series $X(t)$ and $X(t+\tau)$ take the values $s_i$ and $d_i$ simultaneously.
We notice that for independent series $\mathcal{H}(\mathcal{S},\mathcal{D}) = \mathcal{H}(\mathcal{S}) + \mathcal{H}(\mathcal{D})$, so that $\mathcal{I}(\mathcal{S},\mathcal{D}) = 0$, while for the zero delay $\mathcal{I}(\mathcal{S},\mathcal{S}) = \mathcal{H}(\mathcal{S},\mathcal{S}) = \mathcal{H}(\mathcal{S})$.
The method was tested on the Lorenz attractor by Hongre *et al.* (1995) [106] who calculated the mutual information for different time delays and forenoise levels of 0, 1, 5, 10 and 20 percent. One can see on figure 9.7 that for 20,000 samples of the scalar projection of the Lorenz attractor orbit the results are consistent until noise levels of 20 percent. The optimal time lag is 5 samples.

## 9.3.2 The minimum embedding dimension

To answer the second question - what minimum embedding dimension $d_E$ to use- we investigate the behaviour of neighbouring points of the phase-space orbit under changes in the embedding dimension $d_E \to d_E + 1$. When the number of **false nearest neighbours** comes to zero, we suppose that the orbit has been smoothly embedded (unfolded) in dimension $d_E$.
When the embedding dimension is too small, singularities in the attractor result in folding of the orbit, *i.e.* not all points on the orbit which are close to one another will be neighbours because of the dynamics. For the nearest

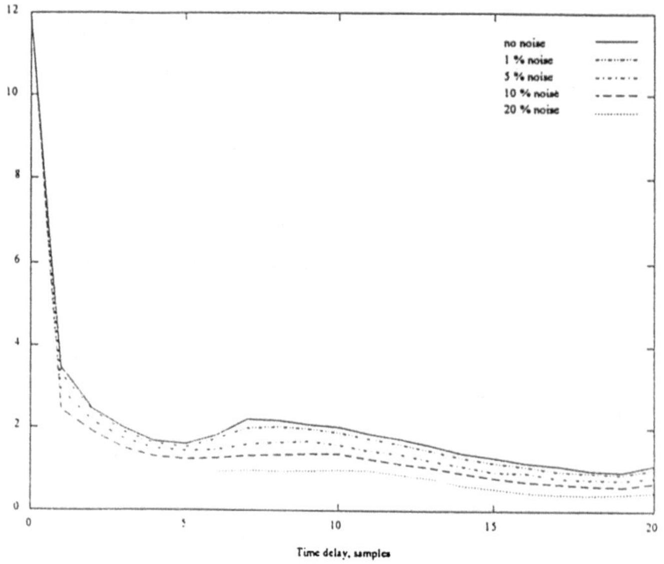


Figure 9.7: *Mutual information method.* The mutual information for different time delays of the Lorenz attractor orbit and noise levels (20,000 samples), (after Hongre *et al.*, 1995).

neighbour $\boldsymbol{y}(t)$ to the point $\boldsymbol{x}(t)$ in $d$ dimensions, the square of the Euclidian distance is

$$R_d^2(t) = \sum_{k=0}^{d-1} (x(t+k\tau) - y(t+k\tau))^2 \ . \tag{9.17}$$

The step in the embedding dimension $d \to d+1$ adds a $(d+1)$th coordinate into each of the vectors $\boldsymbol{x}(t)$ :

$$R_{d+1}^2(t) = R_d^2(t) + (x(t+d\tau) - y(t+d\tau))^2 \ . \tag{9.18}$$

If the increase in distance between $x(t)$ and $y(t)$ is large when going from dimension $d$ to dimension $d+1$ (say, greater than a threshold $R_{d+1}(k) - R_d(k)/R_{d+1}(k) > R_T$), we call $y(t)$ a false nearest neighbour. In practice the value of the threshold seems to be $R_T \approx 10$.

In figure 9.8 we tested the method on a 20,000 samples of a scalar projection of Lorenz attractor orbit with noise levels of 0, 1, 5, 10 and 20 percent. For a noise level of 5 % we still are close to the exact value of embedding dimension $d_E = 3$, and for noise levels 10% and 20% the criterion again gives a reasonable dimension $d_E = 4$.

This example is illustrative since we know the fractal dimension of the orbit is $d_F = 2.06$. Thus, according to Mané (1981) [129] and Takens (1981) [173] theorem, one can smoothly embed the orbit into the space with dimension $d_E > [2d_F] = 4$, which overestimates the real number of degrees of freedom of the Lorenz system.

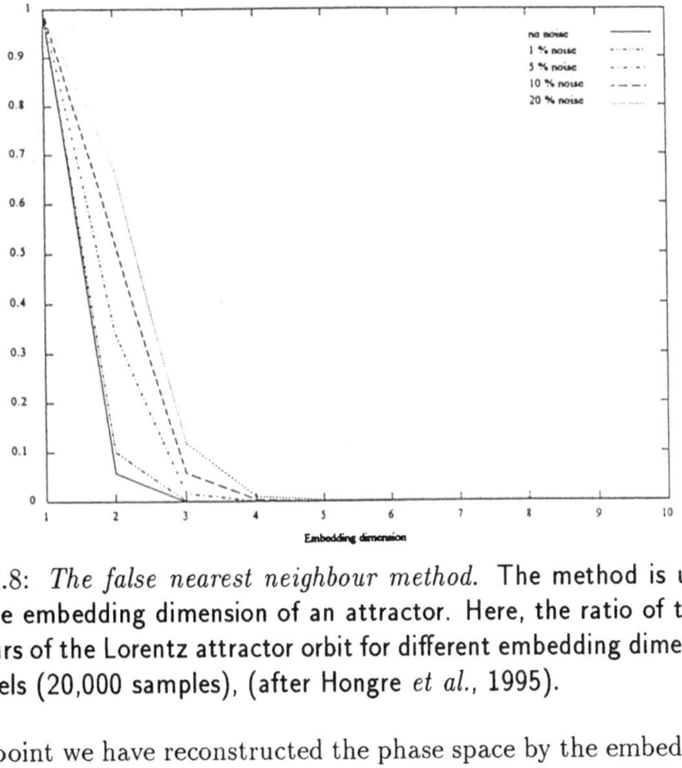


Figure 9.8: *The false nearest neighbour method.* The method is use to calculate the embedding dimension of an attractor. Here, the ratio of the nearest neighbours of the Lorentz attractor orbit for different embedding dimensions and noise levels (20,000 samples), (after Hongre *et al.*, 1995).

At this point we have reconstructed the phase space by the embedding technique and obtained estimates of the topological invariants of the attractor. Now we try to estimate its metric invariants, namely global Lyapunov exponents.

### 9.3.3 The Lyapunov exponents computation (Sano and Sawada method)

Let us consider a small perturbation of the orbit $\delta \boldsymbol{x}\ (0)$ at time $t = 0$ which evolves with the dynamics:

$$\boldsymbol{x}\ (t+1) + \delta \boldsymbol{x}\ (t+1) = \boldsymbol{F}(\boldsymbol{x}\ (t) + \delta \boldsymbol{x}\ (t)). \tag{9.19}$$

It can be approximated locally as

$$\delta \boldsymbol{x}\ (t+1) = \boldsymbol{DF}(\boldsymbol{x}\ (t))\delta \boldsymbol{x}\ (t)\,. \tag{9.20}$$

Here, the Jacobian $\boldsymbol{DF}(\boldsymbol{x}\ (t))$ is considered as a linear operator in the tangent space of $\boldsymbol{x}\ (t)$.
This induction may be rewritten as a composition of such operators along the orbit applied to the initial perturbation

$$\begin{aligned} \delta \boldsymbol{x}\ (t+1) &= \boldsymbol{DF}(\boldsymbol{x}\ (t))\boldsymbol{DF}(\boldsymbol{x}(t-1))\ldots\boldsymbol{DF}(\boldsymbol{x}\ (0))\delta\boldsymbol{x}\ (0) \\ & \boldsymbol{DF}^{(n)}(\boldsymbol{x}\ (0))\delta\boldsymbol{x}\ (t)\,. \end{aligned} \tag{9.21}$$

The Osceledec multiplicative ergodic theorem [139] implies that for a $d$-dimensional ergodic dynamic system there exists a limit operator $\boldsymbol{DF}^{(n)}(\boldsymbol{x}\ (0))$ independent of the initial choice of $\boldsymbol{x}(0)$.
Computation of the Lyapunov spectrum is conducted using an approximation of the tangent flow $\boldsymbol{DF}^{(n)}(\boldsymbol{x}\ (0))$ by a stepwise linear flow map $\boldsymbol{A}_j$which is calculated by a least squares method from a couple of "pseudo-tangent" vectors propagated within a certain interval.

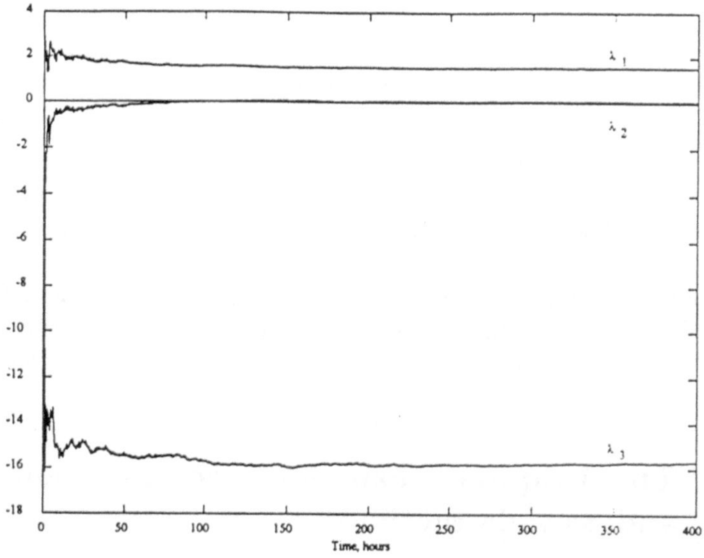


Figure 9.9: *The Lyapunov exponents of Lorenz attractor.* The computation was done on a 20,000 samples series and no noise (after Hongre *et al.* 1995).

For a reference point on the orbit $\boldsymbol{x(j)}$, we call "pseudo-tangent" a set of differences $\boldsymbol{y}^{\,i} = \boldsymbol{x}(j) - \boldsymbol{x}(k_i),\quad i = 1,\ldots,N$ to the $N$ nearest neighbours to

$\boldsymbol{x}(j)$ within a sphere of radius $\epsilon_{\max}$; *i.e.*, $\{\boldsymbol{x}(k_i) \mid \| \boldsymbol{x}(k_i) - \boldsymbol{x}(j) \| \leq \epsilon_{\max}\}$. After a fixed evolution interval $\Delta t = m\tau_s$, the differences $\boldsymbol{y}^i$ have been evolved to the vectors $\boldsymbol{z}^i = \boldsymbol{x}(j+m) - \boldsymbol{x}(k_i+m)$.
The vector $\boldsymbol{z}^i$ should be mapped by $\boldsymbol{z}^i = \boldsymbol{A}_j \boldsymbol{y}^i$. The optimal estimation of the flow $\boldsymbol{A}_j$ is given by the following least-squares ansatz:

$$\min_{\boldsymbol{A}_j} \mathcal{S} = \min_{\boldsymbol{A}_j} \frac{1}{N} \sum_{i=1}^{N} \| \boldsymbol{z}^i - \boldsymbol{A}_j \boldsymbol{y}^i \| . \tag{9.22}$$

Finally one-dimensional Lyapunov exponents are calculated according to

$$\lambda_i = \lim_{n \to \infty} \frac{1}{n\tau_s} \sum_{j=1}^{n} \ln \| \boldsymbol{A}_j \boldsymbol{e}_i(j) \| . \tag{9.23}$$

Results of numerical calculations of the Lyapunov spectrum for the time delay reconstruction of the Lorenz attractor orbit (20,000 samples) are drawn in figure 9.9. The real values of the exponents in this case are $\lambda_1 = 1.5$, $\lambda_2 = 0.0$, $\lambda_3 = -22.5$ (Wolf *et al.* 1985 [180]).

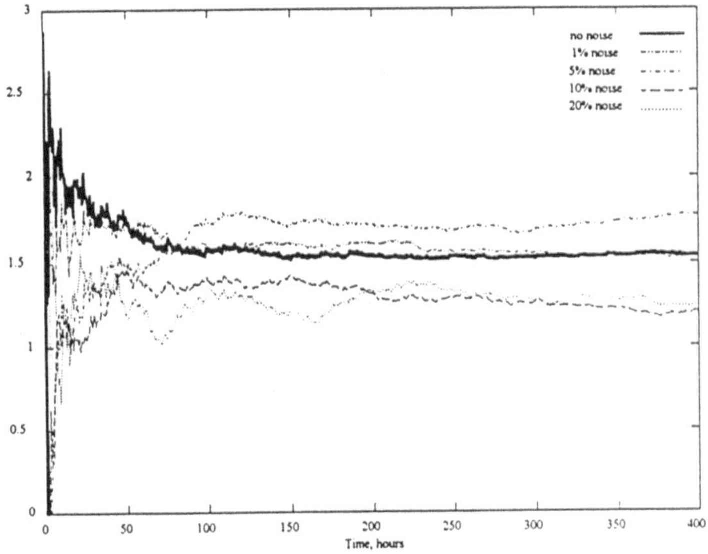


Figure 9.10: *Noise effect on Lyapunov exponents computation.* The effect of noise on the exponents is a decrease of their magnitude especially for high level noise, nevertheless the signs of the Lyapunov spectrum remain the same (Hongre *et al.*, 1995).

Hongre *et al.* (1995) [106] have found a good agreement in the $\lambda_1$ and $\lambda_2$ estimates, although the precision of their programs didn't allow to reach the

very low value of the negative exponent $\lambda_3$. The free parameters of their algorithm were: number of nearest neighbours $N = 20$ in a sphere of radius $\epsilon_{\max} = 0.01$ excluding 50 points on the orbit before and after a reference point, evolution for $\boldsymbol{A}_j$ estimation $m = 4$ $(0.08s)$, GSR reorthogonalization interval $0.3s$ (every 15 points).
They tested also the robustness of the algorithm in the presence of noise (noise levels of 1, 5, 10 and 20 percent). The results are drawn in figure 9.10.
We shall develop in the next book the application of the algorithm to magnetic field series.

### 9.3.4 Calculation of Lyapunov exponents (the method of Wolf *et al.*)

Let us consider a dynamic system which we study through one of its temporal variables $x(t)$, and let us suppose that, from the Grassberger and Procaccia test, the dynamic system is deterministically chaotic, that is to say that it has an attractor in its phase space.
We define the $m$-dimensional phase portrait into which a point of the attractor has coordinates $x(t)$, $x(t+\tau)$, ..., $x(t+(m-1)\tau)$, where $\tau$ is an arbitrary time delay suitably chosen (for exemple by using the mutual information criterion [60] developped below).
In the method which is presented here [180], we localize the nearest neighbours (in the euclidian metrics sense) of the initial point $x(t_0)$, ..., $x(t_0 + (m-1)\tau)$ and we define the distance between the two points as $\mathrm{L}(t_0)$.

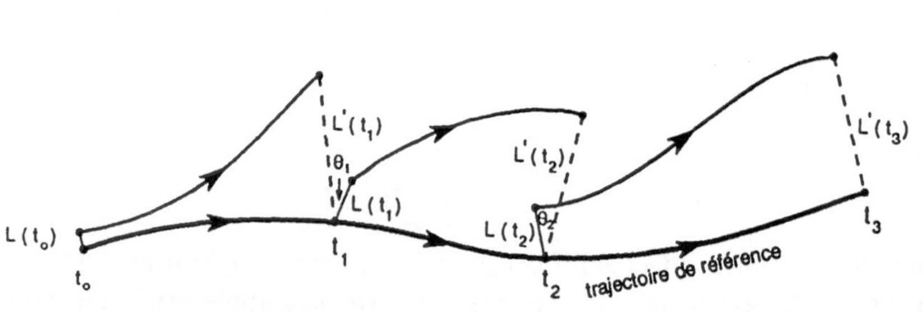


Figure 9.11: *Lyapunov exponents.* Representation in the phase space of the divergence of trajectories starting everywhere from the nearest neighbour of each point of the fiducial trajectory. The increase of average distance permits computation of the bigger Lyapunov exponent (after Wolf *et al.*, 1985).

At time $t_1$ the initial length is now $L'(t_1)$. If we take too long of a time interval $(t_0, t_1)$ we may have $L'(t_1)$ shorter than $L(t_0)$. This may also occur at any place on the attractor where there is folding so that $\lambda_1$ is underestimated. To avoid this, we use a statistical method, all along the series, on $L'(t_k)$. We get

$$\lambda_1 = \frac{1}{t_M - t_0} \sum_{k=1}^{M} \log_2 \frac{L'(k)}{L(k-1)}, \tag{9.24}$$

where $M$ is the total number of steps on the fiducial trajectory. We follow the procedure in figure 9.11.

The fiducial trajectory is representative of the attractor points at times $t_0$, $t_1$, $t_3$, ..., $t_k$. The length $L(t_0)$ is the distance at $t_0$ to its nearest neighbour which, at time $t_1$, changes to the distance $L'(t_1)$; $L(t_1)$ is the distance of the fiducial trajectory point at time $t_1$ to its nearest neighbour (which is not the previous one which is now at a distance $L'(t_1) > L(t_1)$). Next, this second neighbour goes to a distance $L'(t_2)$ at time $t_2$, ..., and so on.

We may again extract more information from the series and we may improve the method when computing the sum of the two largest Lyapunov exponents. The method consists in considering the nearest neighbours at $t_0$, $t_1$, $t_2$, ..., $t_k$ which define areas $A(t_0)$, $A(t_1)$, ..., $A(t_k)$. Then, $A(t_0)$ becomes at time $t_1, A'(t_1)$, while $A(t_1)$ becomes $A'(t_2)$, and so on (see figure 9.12).

Along the whole series we get

$$\lambda_1 + \lambda_2 = \frac{1}{t_M - t_0} \sum_{k=1}^{M} \log_2 \frac{A'(t_k)}{A(t_{k-1})}. \tag{9.25}$$

It is possible to compute from the Lyapunov exponents the dimension of the attractor. Kaplan and Yorke (1978) [112] and Farmer *et al.* (1983) [55] define the Lyapunov dimension which equals, in the case of a 2D mapping, the capacity or similarity dimension. We have

$$d_L = d_C = 1 + \frac{\log \lambda_1}{\log(1/\lambda_2)}. \tag{9.26}$$

It is thus possible to calculate the attractor dimension from Lyapunov exponents.

We shall come back to this property when applying Lyapunov exponents to geomagnetic time series in the third part of this work.

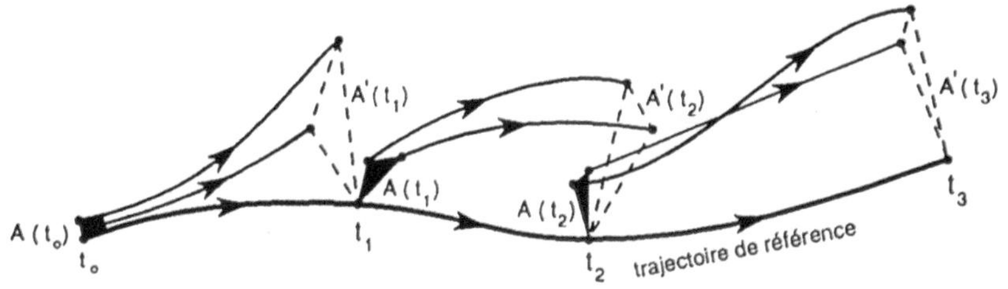


Figure 9.12: *Calculation of the two biggest Lyapunov exponents.* The method of computation uses areas constructed on the two nearest neighbours of each point of the fiducial trajectory (after Wolf *et al.*, 1985).

## 9.4 First Return Mapping

We shall go rapidly through this method which was already developped in chapter 8.

In this method, the flow behaviour is examined through the unidimensional mapping $x_{k+1} = f(x_k)$ which links the coordinate $x_{k+1}$ at the moment $(k+1)T$ to its antecedent coordinate $x_k$ at the moment $kT$. The mapping so defined is called the first return map. The useful information is thus concentrated in one dimension of the phase space and the other ones are not examined.

In the next book on applications, we shall develop an application of this method to the study of sequences of volcanic eruptions.

# Chapter 10

# Bifurcation, Cascades and Chaos

We have introduced the concept of bifurcation when we studied the logistic mapping (8.5). Let us now come back and deeply address this question. Following Ruelle (1994) [156], let us consider a horizontal layer of water heated from below. For small rates of heating we have a conducting regime, and the water remains motionless. For larger rates of heating the fluid starts moving: we are in a convecting regime. In this situation, the intensity of heating is a **bifurcation parameter** and the transition from the conducting to the convecting regime is an example of bifurcation.

## 10.1 Floquet Matrix, Hopf Bifurcation

We have seen that in the simple periodic system, the phase trajectory is tending toward a limit cycle, the Poincaré section of which is reduced to a single point $P_0$. This point is the fixed point of the application

$$\mathcal{T}: \quad P_0 = \mathcal{T}(P_0) = \mathcal{T}^2(P_0) = \cdots \tag{10.1}$$

Let us have a look at the stability of this periodic solution. There are three possibilities for loss of stability. We consider a linear analysis that only retains first-order terms and that will be seen to be sufficient. The application $\mathcal{T}$ is described, to first order, by the *Floquet matrix* $\mathcal{M}$ defined in the vicinity of 0 by

$$\mathcal{M} = [\partial \mathcal{T} / \partial x_i]_{x_i^0} . \tag{10.2}$$

Let us now introduce the properties of the Floquet matrix.

### 10.1.1 Floquet matrix

$P_0$ is the Poincaré section stable point corresponding to a stable limit cycle, while $P_1$, $P_2$, $\cdots$, $P_n$ are points of the Poincaré section that are intersections of the trajectory in the neighbourhood of the limit cycle. We look how $P_2$, $P_3$, $\cdots$, $P_n$ can be obtained from $P_1$.

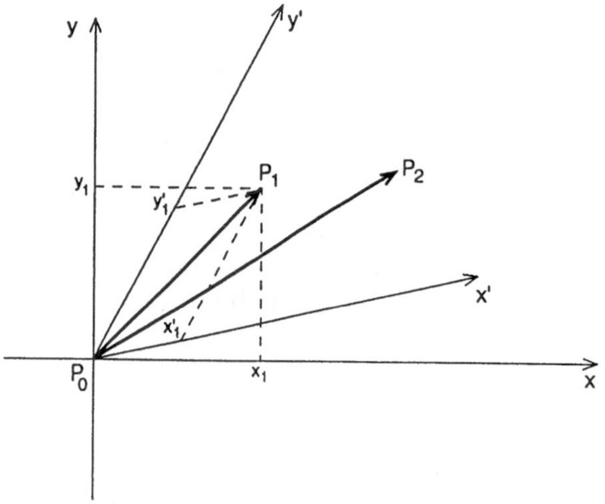


Figure 10.1: *The Floquet matrix.* Here are shown the eigen values axis of the Foquet matrix, $P_0x'$ and $P_0y'$ (after Bergé *et al.*, 1984).

We define a cartesian frame $P_0x$, $P_0y$ in the section plane. The coordinates of the points are $P_1(x_1, y_1)$, $P_2(x_2, y_2)$ $\cdots$ (fig. 10.1).
The linear transformation which defines the $P_2$ coordinates, which are the image of $P_1$ is

$$P_2 = \begin{cases} x_2 & = ax_1 + by_1 \\ y_2 & = cx_1 + dy_1 \end{cases} \tag{10.3}$$

and, in general,

$$P_n = \begin{cases} x_n &= ax_{n-1} + by_{n-1} \\ y_n &= cx_{n-1} + dy_{n-1} \end{cases} \tag{10.4}$$

for which the transformation matrix is

$$\mathcal{M} = \begin{pmatrix} a & b \\ c & d \end{pmatrix}. \tag{10.5}$$

We may write

$$\begin{pmatrix} x_n \\ y_n \end{pmatrix} = \mathcal{M}^{n-1} \begin{pmatrix} x_1 \\ y_1 \end{pmatrix}, \text{or} \quad \overrightarrow{P_0P_n} = \mathcal{M}^{n-1}\overrightarrow{P_2P_1}. \tag{10.6}$$

The computation of $\mathcal{M}^{n-1}$ is not trivial. So to simplify the transform process, we propose to define a new frame $x'y'$ and two numbers $\lambda_1, \lambda_2$ such that $P_2$ is deduced from $P_1$ by,

$$P_2 = \begin{cases} x'_2 &= \lambda_1 x'_1 \\ y'_2 &= \lambda_2 y'_1 \end{cases}, \text{and} \quad P_n = \begin{cases} x'_n &= \lambda_1^{n-1} x'_1 \\ y'_n &= \lambda_2^{n-1} y'_1 \end{cases}. \tag{10.7}$$

The new transformation matrix $\mathcal{M}'$ is diagonal,

$$\mathcal{M}' = \begin{pmatrix} \lambda_1 & 0 \\ 0 & \lambda_2 \end{pmatrix}, \tag{10.8}$$

and,

$$\mathcal{M}'^{(n-1)} = \begin{pmatrix} \lambda_1^{n-1} & 0 \\ 0 & \lambda_2^{n-1} \end{pmatrix}. \tag{10.9}$$

One find the eigenvalues of $\mathcal{M}$, $\lambda_1$ and $\lambda_2$ which are the roots of equation,

$$\left\| \begin{matrix} a - \lambda & b \\ c & d - \lambda \end{matrix} \right\| = 0,$$

$$\lambda^2 - (a + d)\lambda + (ad - bc) = 0. \tag{10.10}$$

The roots may be complex conjugates, $\lambda_1, \lambda_2 = \alpha \pm i\beta$. The eigen vectors $\overrightarrow{V}$ of $\mathcal{M}$ are such that $\mathcal{M} \cdot \overrightarrow{V} = \lambda \cdot \overrightarrow{V}$. They are colinear with axis $P_0x$ et $P_0y$. They are the eigen directions of $\mathcal{M}$.
They are defined, in $xoy$, by

$$ax + by = \lambda x \,,$$

$$cx + dy = \lambda y \,, \tag{10.11}$$

$$\lambda = \lambda_1, \lambda_2 \,.$$

As we have,

$$P_n = \begin{cases} x'_n &= \lambda_1^{n-1} x'_1 \\ y'_n &= \lambda_2^{n-1} y'_1 \end{cases} ,$$

that is the larger modulus of $\lambda$, $\mid \lambda \mid^+$ with respect to 1 which acts in the evolution of $P_n$.
When,

$$\mid \lambda \mid^+ < 1, \quad \lambda^n \to 0, \quad P_n \to P_0 \,,$$

$$\mid \lambda \mid^+ > 1, \quad \lambda^n \to \infty, \quad P_n \to \infty \,. \tag{10.12}$$

The Floquet matrix is such that, after one cycle, the transform $P_1$ of $P_0$ becomes $P_0 + \delta$ and $\mathcal{T}(P_0 + \delta) - P_0 \approx \mathcal{M}\delta$ as $\mid \delta \mid \to 0$.
The stability of the trajectory depends on the Floquet matrix eigenvalues, because, after $m$ cycles $\mathcal{T}^m(P_0 + \delta) - P_0 \approx \mathcal{M}^m \delta$.
The distance will exponentialy tend to $0$, if the eigenvalues of $\mathcal{M}$ remain strictly lower than $1$.
If one of the eigenvalues, at least, is greater than $1$, the distance increases with time and the cycle is unstable.
The important result in relation with Floquet matrix properties may expressed as following: the loss of stability of the limit cycle corresponds to the crossing of the unit circle by at least one Floquet matrix eigenvalue.

### 10.1.2 Hopf bifurcation

Let us now examine the question: How is it possible to move from a periodic attractor to a stange (chaotic) attractor? During the loss of stability of a periodic regime (see above) there are three possible ways.

In figure 10.2, the stability loss is linked to

$$|\overrightarrow{x_0x_1}| < |\overrightarrow{x_0x_2}| < \ldots \tag{10.13}$$

If $\overrightarrow{\delta x_1} = \overrightarrow{x_0x_1}$, and $\overrightarrow{\delta x_2} = \overrightarrow{x_0x_2}$, we get

$$\overrightarrow{\delta x_2} = \mathcal{M}\,\overrightarrow{\delta x_1}\,. \tag{10.14}$$

The Floquet matrix $\mathcal{M}$ has the eigenvalue $\lambda$ .
When,

1. $\lambda = 1 + \epsilon$, $\overrightarrow{\delta x_i}$ is increasing at every cycle,
2. $\lambda = -(1+\epsilon)$, $\overrightarrow{\delta x_i}$ has an increasing modulus, but it points, alternatively, in two opposite directions.
3. $\lambda$ is a complex conjugate, $\lambda = \alpha \pm i\beta$ with $|\lambda| > 1$, the successive $\overrightarrow{\delta x_i}$ rotate with an angle $\gamma$ at each cycle and the $\delta x_i$ modulus increases (fig. 10.2).

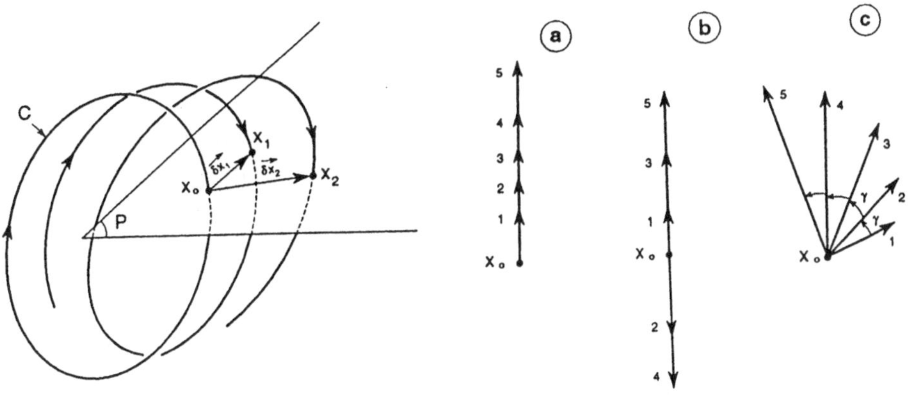


Figure 10.2: *Poincaré section of a phase trajectory.* $C$ is the limit cycle which crosses the section plane at $x_0$. In the Poincaré section plane there are three possible cases, for $\lambda > 1$; $\lambda < -1$; $\lambda = \alpha \pm i\beta$ and $|\lambda| > 1$ (after Bergé *et al.*, 1984).

In the last case, the limit cycle which is defined, on the Poincaré section, by the fixed point $x_0$ is transformed by a Hopf bifurcation, in attractor torus

$\mathcal{T}^2$, the Poincaré section of which is formed by points $T_i$ on a circle, in the case where $|\overrightarrow{\delta x_i}|$ has a finite value. It requires that the non-linear effects, which were neglected in the Floquet matrix transformation, act in order to limit the amplification of $|\overrightarrow{\delta x_i}|$. The figure 10.3 [20] represents this passing of $x_0$, limit cycle section, by instability, through the torus $\mathcal{T}^2$ the section of which is a circle geometric spot of the $T_i$.

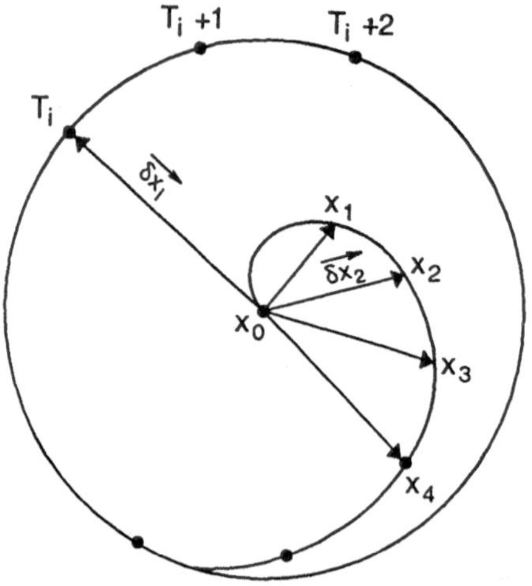


Figure 10.3: *Poincaré section in Hopf bifurcation.* One passes from $x_0$, intersection of the limit cycle, to the circle of the $T_i$, intersection of the tore $T^2$ (after Bergé *et al.*, 1984).

### 10.1.3 Passing from torus $\mathcal{T}^2$ to torus $\mathcal{T}^3$. Ruelle-Takens theory

The following item will be qualitative. Fro better developments, see Ruelle and Takens (1971) [153]. In this study let us consider a viscous laminar flow. If the control parameter (here the Reynolds number) increases, the regime losses its stability and becomes oscillating with period $f_1$. When the process is repeated by increasing the control parameter, one may imagine that a series of Hopf bifurcations may occur implying appearance of independant frequencies $f_2$, $f_3$.

The Ruelle-Takens theory claims (and proves) that the torus $\mathcal{T}^3$ thus obtained may become unstable and may be replaced by a strange attractor.

In figure 10.4I of the frequency spectrum and on the diagram showing the system evolution as a function of the control parameter (fig. 10.4II) we may observe successive steps from a periodic regime (fig.10.4 Ia) to a two-frequency quasi-periodic regime (fig. 10.4Ib), and, finally, to the chaotic regime.

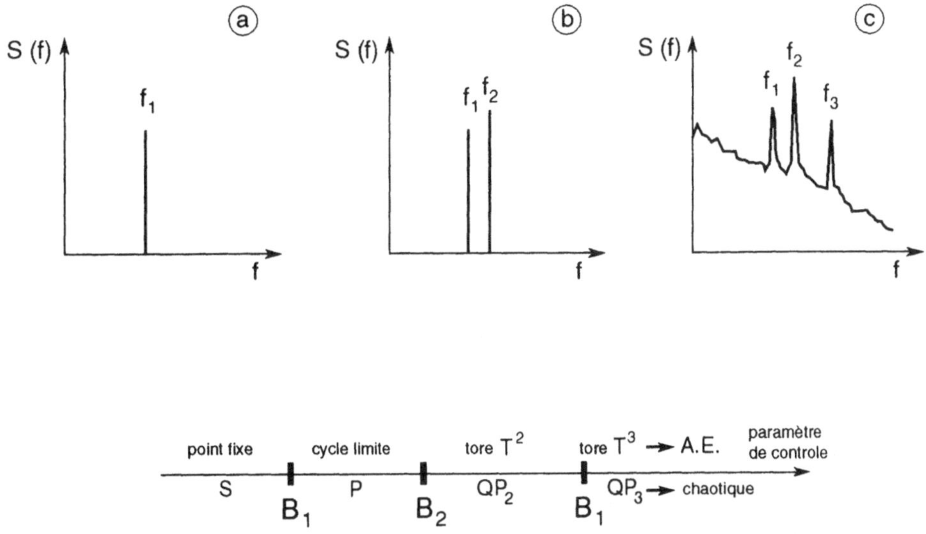


Figure 10.4: *Ruelle and Takens theory.* One may observe successive ways from tore $\mathcal{T}^2$ to tore $\mathcal{T}^3$ quasi-periodic, then chaotic. The diagram, where $B_1, B_2, B_3$ are the successive bifurcations, summarizes the different steps (after Bergé *et al.*, 1984).

The Ruelle and Takens proof goes further and shows that the torus $\mathcal{T}^3$ may become completely destroyed and replaced by a strange attractor which is stable and doesn't change whenever submitted to new perturbations.

Unlike the earlier Landau and Lifschitz (1971) theory [121], Ruelle and Takens have shown that a system with a small number of degrees of freedom, might generate a chaotic regime. It is possible to find more simple models also leading to chaos, one of them starting from the torus $\mathcal{T}^2$.

## 10.2 Curry and Yorke Model and Route to Chaos

Instead of passing to chaos by a Tore $T^3$ transformation in a strange attractor, the Curry and Yorke model (1977) [39] passes directly from a two frequency quasi-periodicity to chaos. As they claim in their famous paper: "We feel our objective of investigating the transition to chaos via Hopf bifurcation is quite modest but is interesting and perhaps illuminating".

The study of the route to chaos is not rigorous. It was performed numericaly from a Poincaré section of a three-dimensional flow. To get chaos, the Curry and Yorke numerical investigation considers mappings which are compositions of two simpler homeomorphisms. To obtain chaos the total mapping has to be non-linear, involve stretching and folding, and to present a Hopf bifurcation.

The model result is a composition of two mapping functions:

$$\Im = \Im_1 \circ \Im_2 \,, \tag{10.15}$$

where $\circ$ means: "followed by".

The mapping $\Im_1$ is given by the following relationships, in polar coordinates, between iterates $k+1$ and $k$,

$$\Im_1 = \begin{cases} \rho_{k+1} & = \epsilon \log(1+\rho_k) \\ \theta_{k+1} & = \theta_k + \theta_0 \end{cases} \tag{10.16}$$

where $\theta_0$ positive is fixed. $\epsilon \geq 1$ is the control parameter.

The mapping

$$\Im_2 = \begin{cases} X_{k+1} & = X_k \\ Y_{k+1} & = Y_k + X_k^2 \end{cases} \tag{10.17}$$

is given in cartesian coordinates. Note the similarity between $\Im_2$ and the Henon model mapping (see 9.1).

Figures 10.5 a and b are showing the result of the mapping $\rho_{k+1} = \epsilon \log(1+\rho_k)$, according to wether the value of $\epsilon$ is less than, or greater than $1$.

When $\epsilon < 1$, $\rho$ is tending towards $0$, when $\epsilon > 1$, $\rho$ is tending towards a limit value $\rho_L$. In 10.5c, the mapping $\Im_1$, $\theta_{k+1} = \theta_k + \theta_0$ is presented which corresponds to a rotation of $\theta_0$ so that, generally, the result is the attractor circle $C$.

We have fulfilled the two initial conditions, dissipation and Hopf bifurcation. Indeed, when passing through the value $\epsilon = 1$ of the contol parameter, bifurcation occurs, from the limit cycle, to an invariant tore.

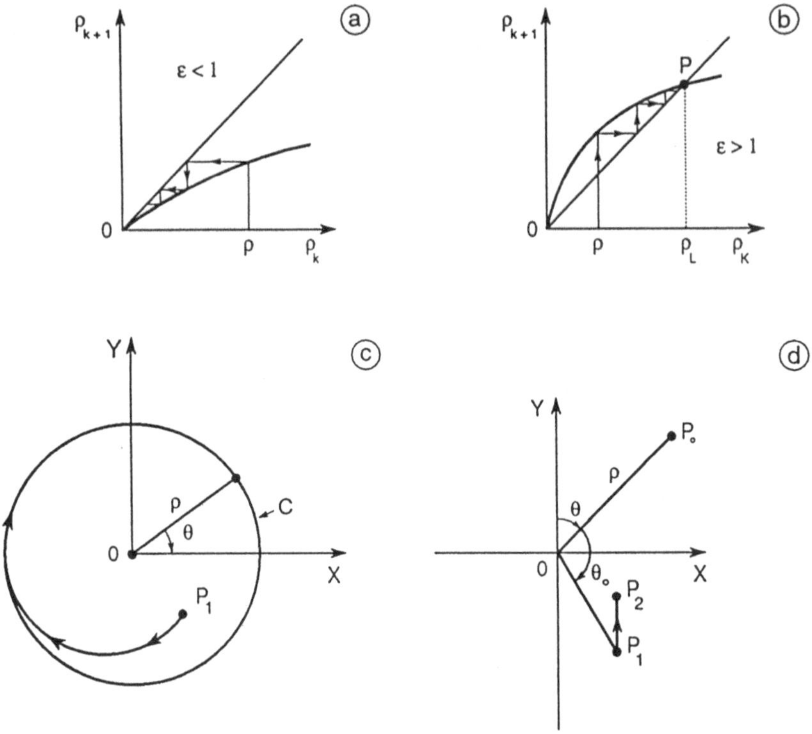


Figure 10.5: *Curry and Yorke model.* The successive graphs show the successive mappings, their resulting point when $\epsilon > 1$ and the passing from point $P_0$ through $P_1$, to $P_2$, after the mapping $\Im$ (after Curry and Yorke, 1977).

The last condition of non-linearity is obtained by the mapping $\Im_2$, $Y_{k+1} = Y_k + X_k^2$, which breaks the coupling between $\rho$ and $\theta$.
On figure 10.5d, the mapping $\Im$ permits passing from $P_0$ to $P_2$, by first passing through $P_1$ (mapping $\Im_1$) and then passing from $P_1$ to $P_2$ (mapping $\Im_2$).
In this mapping, after a certain number of transitory iterates (a point was chosen, and iterated one hundred times), we obtain in the plane of the Poincaré section, the set of representative points of the attractor (or rather its section by $xoy$).
Figures 10.6 show some examples when $\theta_0 = 2$ radians and $\epsilon = 1.27$ then, $\epsilon = 1.48$. For $\epsilon = 1.27$, we have a quasiperiodic regime with two frequencies $f_1$ and $f_2$, where the curve, on the section, is a torus $T^2$.
Beyond a critical value $\epsilon_c = 1.3953$ (for example, when $\epsilon = 1.48$), there

appear asperities (an infinite number) on the section which seems to get thicker. Their dimension is greater than 1 (continuous line of the circle) and is therefore fractal. This means that the torus $\mathcal{T}^2$ has been broken when $\epsilon \geq \epsilon_2$ and that the attractor is becoming strange.

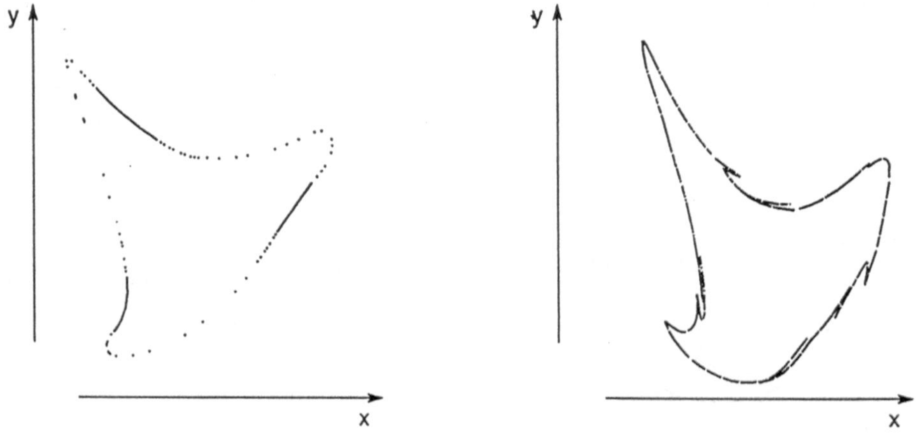


Figure 10.6: *Results of the Curry and Yorke mapping.* For $\theta_0 = 2$ radians and for $\epsilon = 1,30$ (on the left side) and $\epsilon = 1,45$ (on the right side) (after Curry et Yorke, 1977).

One observes easily, by taking neighbouring points on the section, that after some iterates, they become separated (by I.C.S., the strange attractor property which was developped previously) and then return to being close to each other. Stretching and folding are necessary and sufficient conditions for the existence of chaos.

The diagram of figure 10.7 summarizes the different steps of the route to chaos, in the Curry and Yorke model.

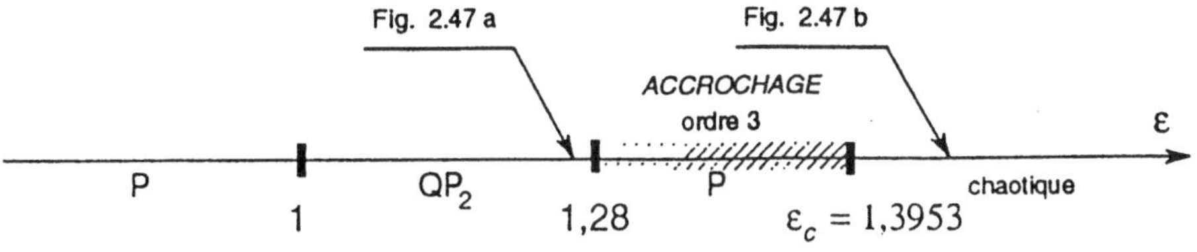


Figure 10.7: *The route to chaos.* Summary of the successive steps in passing to chaos, in the Curry and Yorke model, when $\epsilon$ increases.

## 10.3 Definition of Chaos

As we have already seen, there is no precise definition of chaos. Following Falconer (1990) we may assess that $f$ would certainly be regarded as chaotic on $F$ if the following are all true.

1. The orbit $\{f^k(x)\}$ is dense in $F$.

2. The periodic points of $f$ in $F$ (points for which $f^p(x) = x$ for some positive integer $p$) are dense in $F$.

3. $f$ has *sensitive dependance on initial conditions*; that is, there is a number $\delta > 0$ such that for any $x$ in $F$ arbitrarily close to $x$ such that $| f^k(x) - f^k(y) | \geq \delta$ for some $k$. Thus points that are initially close to each other do not remain close under iterates of $f$.

## 10.4 Control of Chaos

The sensitive dependance on initial conditions allows control of a chaotic dynamic system, even though this is impossible with a stable system. Indeed, a very tiny perturbation (the butterfly effect) can be used to change (i.e., to stabilize) dynamic behaviours of the system. Control of chaos is based on these properties. It consists in directing chaotic trajectories in the phase space to a desired state. We present here a method of stabilizing chaotic

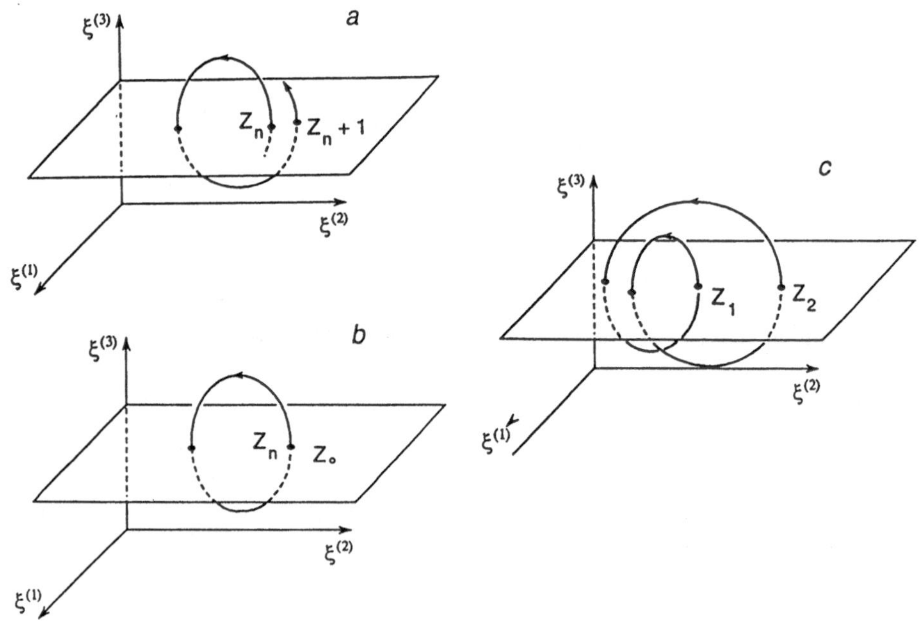


Figure 10.8: *Some Poincaré surface-of-sections for different orbits.* On the Poincaré surface-of sectionte $\xi_3$ = constant we observ the differente trajectories period-1 orbit in b, periode-2 in c, general in a, (Shinbrot *et al.*, 1993).

trajectories which was proposed by Shinbrot *et al.*(1993) [163]. One of the properties of a chaotic system is that there exists a steady state and an infinity of different unstable periodic orbits embedded in chaotic motion. These orbits offer a great potential advantage if one wants to control a chaotic system.

The strategy adopted by the authors was the following. First, they examine the unstable steady states and low-period unstable periodic orbits embedded in chaotic motion. For each of those unstable orbits, they ask whether the system performance would be improved if that orbit were actually followed. They then select one of the unstable orbits that yields improved performance. Assuming the motion of the free-running chaotic orbit to be ergodic, eventually the chaotic wandering of an orbit trajectory will bring it close to the chosen unstable periodic orbit or steady state. When this occurs, they can apply their small controlling perturbations to direct the orbit to the desired periodic motion or steady state.

As a simple illustrative example of controling a chaotic system Shinbrot *et al.* (1993) consider the *Hénon map* (see section 9.1).
The Hénon mapping is two-dimensional where the system state at time $n$ (where $n = 0,\ 1,\ 2,\ \ldots$) is given by two scalar variable, $x_n$ and $y_n$. The map specifies a rule for evolving the state of the system at time $n$, to the state at time $n+1$. Let us recall the formulation of the Hénon map,

$$\begin{aligned} x_{n+1} &= p + 0,3y_n - x_n^2\,, \\ y_{n+1} &= x_n\,, \end{aligned} \tag{10.18}$$

where $p$ is the control parameter.
The value, $p_0 = 1.4$ is given. Thus, given an initial state $(x_1, y_1)$, we may calculate $(x_2, y_2)$, then $(x_3, y_3)$, and so on.
Iterations converge to a strange attractor (*cf.* section 9.1). We recall that its fractal dimension is $1.26$. In figure 10.8 are shown some sections with unstable periodic orbits: a period-1 point, this point $A_\star$ is revisited every map iteration; period-2 points, $B_1$ and $B_2$ are revisited alternatively ($B_1 \to B_2 \to B_1 \to B_2 \cdots$); and period-4 points, $C_1,\ \ldots,\ C_4$ are cycled every 4 iterations.
As we have a mapping, the time is given in a discrete form. In many problems involving dynamic systems we generally find systems implying $M$ first-order, autonomous ordinary differential equations, like $d\xi/dt = \mathbf{G}(\xi)$, where $\xi(t) = (\xi^{(1)}(t),\ \xi^{(2)}(t),\ \ldots, \xi^{(M)}(t))$ is an $M$-vector, and the continuous variable $t$ denotes time.
In such a case, discrete time systems may be useful as they may be reduced by the Poincaré surface-of-section technique. The $M$-dimensional continuous time system is reduced to the $M-1$ dimensional map. This operation is shown in figure 10.8 for $M = 3$. The continuous time trajectory is drawn as well as the corresponding discrete time points on the section $\mathbf{Z}_1,\ \mathbf{Z}_2,\ \ldots,$ where $\mathbf{Z}_n$ represents the $M-1$ coordinate vector (here 2 coordinates) specifying the position on the section of the $n$th upward piercing of the surface. Given a $\mathbf{Z}_n$, we can integrate the equation $d\xi/dt = \mathbf{G}(\xi)$ forward in time from that point, until the next upward piercing of the section surface, at the surface coordinate $\mathbf{Z}_{n+1}$. Thus, $\mathbf{Z}_{n+1}$ is uniquely determined by $\mathbf{Z}_n$ and there must exist a map, $\mathbf{Z}_{n+1} = \mathbf{F}(\mathbf{Z}_n)$, which permit passing from a point of the trajectory on the section to the next
Even if it is not possible to explicitely determine $\mathbf{F}$, the fact that it exists is useful. Figure 10.8 shows a period-1 orbit which corresponds to a map on the Poincaré surface-of-section, $\mathbf{Z}_\star = \mathbf{F}(\mathbf{Z}_\star)$. In the same figure, a period-2 orbit gives, on the Poincaré surface-of-section, a map $\mathbf{Z}_2 = \mathbf{F}(\mathbf{Z}_1),\ \mathbf{Z}_1 = \mathbf{F}(\mathbf{Z}_2)$.

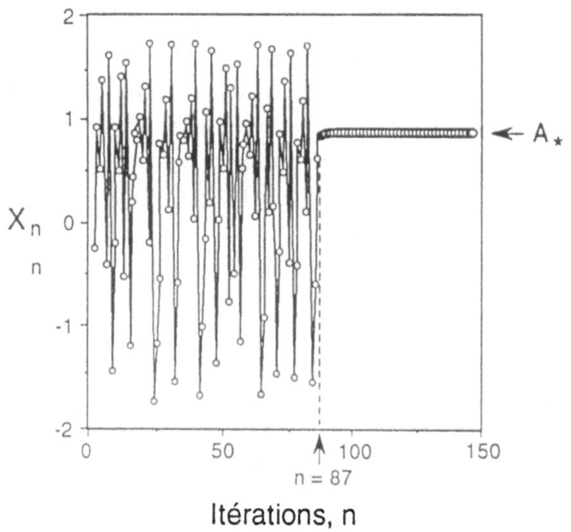


Figure 10.9: *The control of chaos.* Example of stabilisation in a period-1 state for the Hénon map (Shinbrot *et al.*, 1993).

Let us now look at the general case of an unstable periodic orbit. To simplify, one considers a period-1 orbit of some $N$-dimensional map $\mathbf{Z}_{n+1} = \mathbf{F}(\mathbf{Z}_n,\ p)$, where $\mathbf{Z}$ is a $N$-dimensional vector and $p$ is a control parameter. Let us start with an approximation of the dynamics near the period-1 point, denoted $\mathbf{Z}_\star$ (where $\mathbf{Z}_\star = \mathbf{F}(\mathbf{Z}_\star, p_0)$), for the values of the parameter $p$ close to $p_0$, and involving the linear map,

$$(\mathbf{Z}_{n+1} - \mathbf{Z}_\star) = \mathbf{A} \cdot (\mathbf{Z}_n - \mathbf{Z}_\star) + \mathbf{B}(p - p_0), \tag{10.19}$$

Here, $\mathbf{A}$ is a $N \times N$ dimensional Jacobian matrix (*cf.* III) and $\mathbf{B}$ is an $N$-dimensional column vector, where $\mathbf{A} = \partial \mathbf{F}/\partial \mathbf{Z}$ and $\mathbf{B} = \partial \mathbf{F}/\partial p$ and where these partial derivatives are evaluated at $\mathbf{Z} = \mathbf{Z}_\star$ and $p = p_0$.
Let us assume now that we can adjust the parameter $p$ on each iteration. One determines $\mathbf{Z}_n$ and on that basis one proceeds to a little change in $p$ from the nominal value $p_0$. One then replaces $p$ by $p_n$. Considering the relation as linear, one gets,

$$(p_n - p_0) = -\mathbf{K}^T \cdot (\mathbf{Z}_n - \mathbf{Z}_\star), \tag{10.20}$$

where $\mathbf{K}$ is a constant $N$-dimensional column vector and $\mathbf{K}^T$ its transpose. The choice of vector $\mathbf{K}$ specifies the control law of $p_n$ for each iteration. After substitution of equation 10.20 in equation 10.19 we obtain,

$$\delta \mathbf{Z}_{n+1} = (\mathbf{A} - \mathbf{B} \cdot \mathbf{K}^T)\delta \mathbf{Z}_n , \tag{10.21}$$

where $\delta \mathbf{Z}_n = \mathbf{Z}_n - \mathbf{Z}_\star$.
Thus, the period-1 point, $\mathbf{Z}_\star$, will be stable if one can choose $\mathbf{K}$ so that the matrix $(\mathbf{A} - \mathbf{B} \cdot \mathbf{K}^T)$ only has eigenvalues with modulus smaller than unity so that we have $\delta \mathbf{Z}_n \to 0$ (that is to say, $\mathbf{Z}_n \to \mathbf{Z}_\star$) as $n \to \infty$. The choice of $\mathbf{K}$ constitutes the standard problem in the theory of control of chaos.
This procedure gives the time-dependent parameterd values, $p_n$, required to stabilize an unstable period-1 point in a chaotic system. One considers that according to practical constraints one cannot make too large the deviations of $p_n$ from $p_0$. Thus, one considers that $\mid p_n - p_0 \mid$ is bounded by some maximum value, $\delta p_{max}$. Thus, equation 10.20 gives, $\delta p_{max} > \mid \mathbf{K}^T \cdot (\mathbf{Z}_n - \mathbf{Z}_\star) \mid$. If the system state is outside this region, one applies no perturbation and waits until the state falls within the given region. On applies then the tiny perturbation which is appropriate to satisfy equation 10.20.

To illustrate this procedure, the authors show the results of stabilization of a periodic orbit through $\mathbf{A}_\star$ on the Hénon attractor, by adjustment of the $p$ value by less than 1 % of its nominal value. One starts from a random initial point, on the attractor, then one observes (fig. 10.9) that for the first 86 iterations, the trajectory moves chaotically on the attractor, never falling within the desired region about $\mathbf{A}_\star$. Then on the 87th iteration, the system state falls within the desired region, and thereafter is held near to $\mathbf{A}_\star$.
Figure 10.10 shows how, when a periodic orbit is known, the matrix $\mathbf{A}$ and vector $\mathbf{B}$ may be obtained from trajectory observations on the attractor. Let us assume that we have collected a long data string of observed surface-of-section piercings, $\mathbf{Z}_1$, $\mathbf{Z}_2$, and so on. If two successive $\mathbf{Z}_s$ are close to each other, say $\mathbf{Z}_{100}$ and $\mathbf{Z}_{101}$ then there will typically be a period-1 orbit $\mathbf{Z}_\star$ nearby (fig. 10.10). Having observed such a close first return we then search the succeding data for other close-return pairs $(\mathbf{Z}_n, \mathbf{Z}_{n+1})$ restricted to the small area of the original first return. One may find many such pairs if the data set is long enough.

Assuming that the region is small, one will try to fit these close returning pairs with a linear relation,

$$\mathbf{Z}_{n+1} = \hat{\mathbf{A}} \cdot \mathbf{Z}_n + \hat{\mathbf{C}} . \tag{10.22}$$

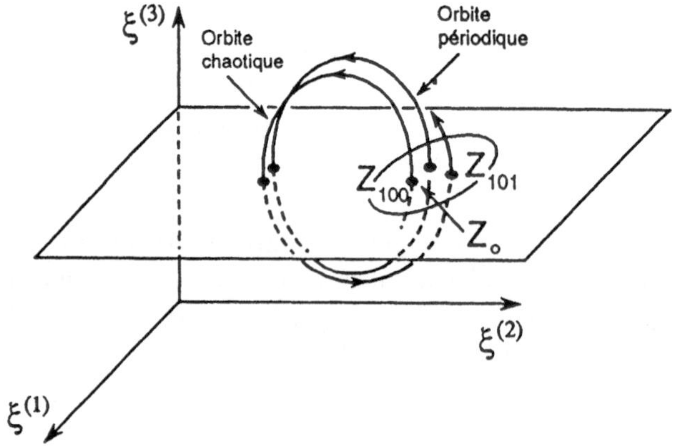


Figure 10.10: *Searching a stable period-1 orbit.* That consists in finding a point $\mathbf{Z}_\star$, from a long data string $\mathbf{Z}_{100}$, $\mathbf{Z}_{101}, \ldots$ (Shinbrot *et al.*, 1993).

Generally if there is noise in the data set, one would want to use as many pairs as possible and fit the matrix $\hat{\mathbf{A}}$ and the vector $\hat{\mathbf{C}}$ to the data using a least-squares criterion. Thus $\hat{\mathbf{A}}$ is an approximation to the Jacobian matrix $\mathbf{A}$ of equation 10.19, and the period-1 point, $\mathbf{Z}_\star(p)$, is approximated by $(1 - \hat{\mathbf{A}})^{-1} \cdot \mathbf{C}$. To find $\mathbf{B}$ of equation 10.19, one changes $p$ slightly ($p \to p + \Delta p$) redetermines the period-1 point ($\mathbf{Z}_\star(p + \Delta p)$, as before) and finally approximates $\mathbf{B}$ as $[\mathbf{Z}_\star(p + \Delta p) - \mathbf{Z}_\star(p)]/\Delta p$.
To find period-2 orbits, one proceeds in the same way, but for pairs $(\mathbf{Z}_n, \mathbf{Z}_{n+2})$ that are close after two surface-of-section piercings, and so on for period-4 orbits and higher periods.

# Chapter 11

# Self Organisation

The concept of self-organized criticality (SOC) has been suggested to be relevant for understanding temporal and spatial scaling in a wide class of dissipative systems with extended degrees of freedom. The nature of this SOC have led Bak, Tang and Wiesenfeld (1987, 1988) [9, 10] to suggest that it may account for the ubiquitous presence in nature of scale-invariant phenomena such as 1/f noise and fractal structures. The concept was proposed as an attempt to explain the existence of self similarties over extended ranges of spatial and temporal scales. It proposes that a spatio-temporal nonlinear dynamic system, with quasi-static incoming and outcoming fluxes, evolve spontaneously toward a stationary self-organized critical state with no length or time scales others than those deduced from the size of the system and the elementary cell.

## 11.1 The Evolving Sandpile Experiment

One system which has been studied extensively for self-organized criticality is the theoretical sandpile [9, 10, 111, 108, 98]. We describe here the last one [98] dealing with the experimental study of critical-mass fluctuations in an evolving sandpile. In this experiment the sandpile is built up to a steady state and then subsequently perturbed by the addition of single grains of sand.

The experimental apparatus is illustrated in Fig. 11.1

According to the scheme of figure 11.1, when a change in mass comparable to a grain of sand is detected, the computer stops the rotation of the funnel, and thus the flow of sand, until any avalanches have occured and the mass of the pile has stabilized.

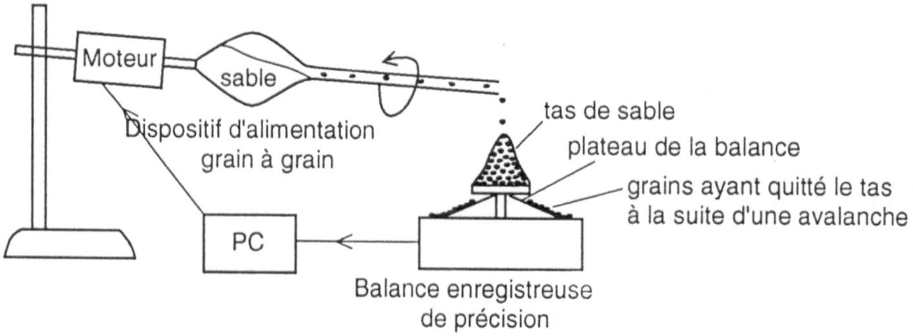


Figure 11.1: *The evolving sandpile experiment.* The sand grains fall out of the tube. They land near the top of a sandpile which is built on a circular disk, which, inturn, is supported on the weighing pan of an analytic balance. The balance is interfaced to an IBM PC which monitors the mass of the sandpile (see Held *et al.*, 1990).

Thus the mass of the sandpile is obtained as a discrete function of the number of grains dropped onto it. Irregularities in the dropping rate, as well as variations in the durations of the avalanches, are effectively removed from the data.

In the Held *et al.* 's experiment, the accuracy of the balance was $\pm 0.0001g$ while its capacity was $100g$. The average mass of a grain was $0.0006g$, and the sandpile base diameter was $4cm$ while its mass was about $15g$. The average time between dropping events was approximately $10s$. The experimental data were collected for about 30,000 dropping events.

**The results**

In figure 11.2, are plotted the mass of the sandpile as a function of the number of grains dropped onto it. The scaling invariance appears on the different graphs.The second graph is a $15\times$ magnification of the small box of the first graph. In the third we magnify $20\times$ the small box of the second graph so that on the last scale, changes in masses are corresponding to single grains of sand. We observe that the mass exhibits fluctuations over periods ranging from one to several thousand dropping events, with avalanches ranging between one and several hundred grains of sand.

To quantify these fluctuations the authors plot the probability density $P(M)$

of an avalanche of mass $M$ for different diameters of sandpiles and for all of these the probability of an avalanche falls off monotonically with increazing avalanche size. For example, for a 4-$cm$-diameter pan, the falloff is approximately a power law for avalanches between 3 and 80 grains ($0.002g$ and $0.05g$) : $P(M) \sim M^{-2.5}$.

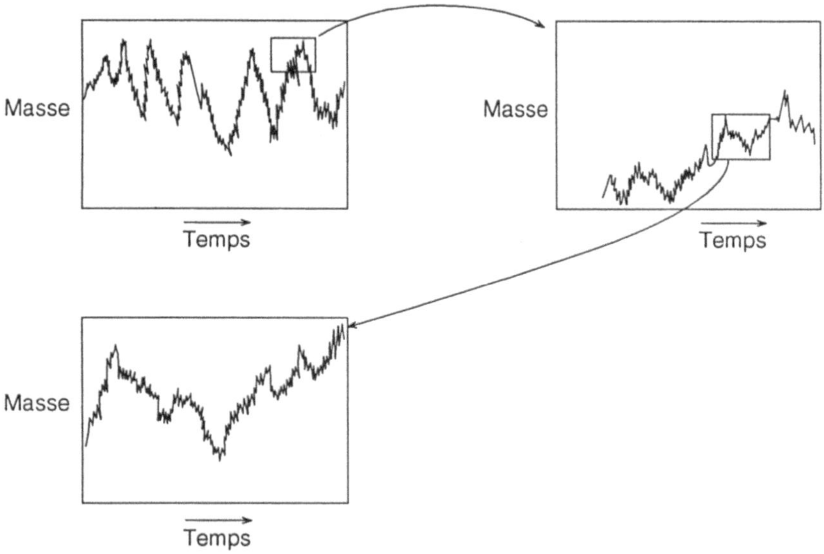


Figure 11.2: *Variations of mass as a function of grains dropped onto the sandpile.* The three last graphs show successively magnification of the box of the previous one (see text ; after Held *et al.*, 1990).

With increasing pan diameter, the probability of large avalanches increases and that of the small avalanches decreases. If the system exhibits scale-invariant fluctuations (similar to those observed at the critical point of a second-order phase transition) we would expect the distribution of avalanches to show finite-size scaling :

$$P(M, L) = (1/L^{\beta})g(M/L^{\nu}), \tag{11.1}$$

where $P(M, L)$ is the probability of an avalanche of mass $M$ for a sandpile of base diameter $L$ and $g$ is a universal function. When rescaling the data, the authors have found that in order to get the rescaled avalanche distributions for different size sandpiles to lye almost exactly on the same universal curve, they must use $\beta = 2\nu = 1.8$. This indicates that the data shows finite-size scaling. Furthermore, figures 11.2 demonstrate that the sandpiles are in a self-organized critical state.

### The Models

We present, firstly, the Hwa and Kardar model [108]. They assume that the system achieves a steady state which is on average a flat surface (the surface of the sandpile). The surface fluctuations are now described by a function $h(\mathbf{x}, t)$, that measures deviations from the average steady-state surface as shown on the figure 11.3. The component of gravity parallel to the surface creates a direction of transport $\hat{\mathbf{T}}$. Then the authors define $\mathbf{x}_{\|} \equiv (\hat{\mathbf{T}} \cdot \mathbf{x})\hat{\mathbf{T}}$ and $\mathbf{x}_{\perp} \equiv \mathbf{x} - \mathbf{x}_{\|}$;

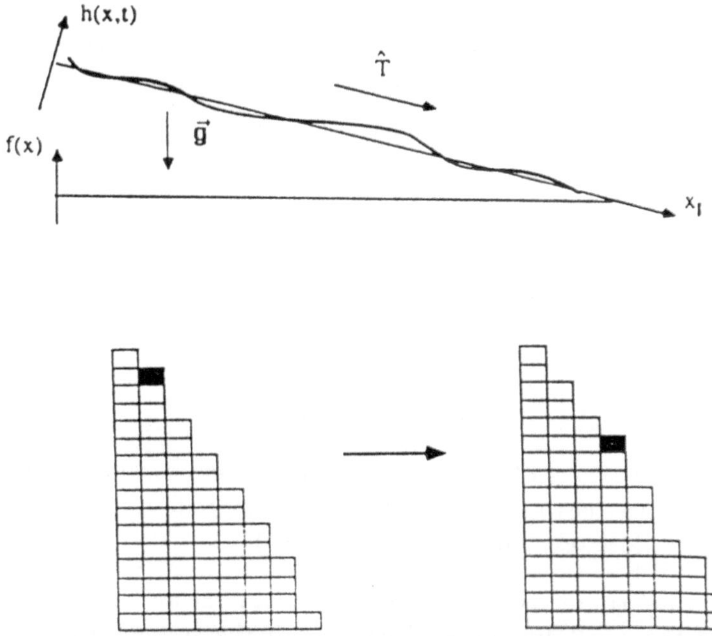


Figure 11.3: *On the surface of the sandpile.* The height function is the deviation from the flat steady-state sans profile. The arrow indicate the transport direction. $\hat{\mathbf{T}}$ is the transport direction (after Hwa and Kardar, 1989).

Considering symmetry and conservation laws in the system they give the simplest equation of motion having non-linearity.

$$\frac{\partial h(\mathbf{x},t)}{\partial t} = \nu_{\|}\partial_{\|}^2 h + \nu_{\perp}\nabla_{\perp}^2 h - \frac{\lambda}{2}\partial_{\|} h^2 + \eta(\mathbf{x},t)\,, \tag{11.2}$$

where the first term is the relaxation of the height through surface tension, the third is due to absence of $x_{\|} \to -x_{\|}$ symmetry and hence related to the transport, while $\eta$ is a stochastic noise.

Equation 11.2 is a driven-diffusion equation. When $\eta = 0$, there is a local height-conservation law and a transport current $\mathbf{j}(\mathbf{x}, t)$, satisfying $\partial_t h + \nabla \cdot \mathbf{j} = 0$, where

$$\mathbf{j} = -\nu_\perp \nabla_\perp h - \nu_\| \partial_\| h \hat{\mathbf{T}} + (\lambda/2) h^2 \hat{\mathbf{T}}.$$

The non-linearity here represents an analytic term trying to mimic the discrete step function. The addition of sand grains from outside destroys the local height-conservation rule.

To summarize the model we point out that a simple continuum equation is constructed to describe fluctuations around a steady state in a flowing sandpile. The principle of scale invariance and self-similarity is understood in terms of a conservation law in dynamics. Finally, the authors use a dynamic renormalization-group calculation to determine various critical exponents in all dimensions.

## 11.2 General Aspects of Self-Organized Criticality

Many natural dynamic systems present the caracteristics of SOC, for example, sandpiles, snow avalanches, "game of life", volcanic eruptions, lithospheric blocks and eathquakes swarms. some of them will be examined in the following book on applications in Physics of the Earth.

This concept of SOC proposes that a spatio-temporal non-linear dynamic system, with quasi-static incoming fluxes localized, for instance, at the borders, evolve spontaneously toward a stationary self-organized critical state with no length or time scales others than those deduced from the size of the system and that of the elementary cell ([11, 167, 169].

Systems as large and as complicated as the earth's crust and the echosystem can break down not only under the force of a mighty blow but also at the drop of a pin [11]. Large interactive systems perpetually organize themselves to a critical state in which a minor event starts a chain reaction that can lead to a catastrophe.

Following Sornette (1993) [167], we can give a formulation of the sandpile model in terms of diffusion equation.

Let $\varphi(\mathbf{r}, t)$ the local height of the sandpile to the point $\mathbf{r}$ at time $t$.

The instability condition is $\varphi(\mathbf{r}, t) \geq \varphi_s$ where $\varphi_s$ is the threshold and its transmission to neighbours can be expressed mathematically under the following form:

$$\varphi(x,y,t+1) = \varphi(x,y,t) - \theta[\varphi(x,y,t) - \varphi_s]$$
$$+(1/4)\theta[\varphi(x+1,y,t) - \varphi_s] + (1/4)\theta[\varphi(x-1,y,t) - \varphi_s]$$
$$+(1/4)\theta[\varphi(x,y+1,t) - \varphi_s] + (1/4)\theta[\varphi(x,y-1,t) - \varphi_s]$$

where $\theta[X]$ is the Heaviside function equal to 1 when $X > 0$ and to 0 when $X \leq 0$. When introducing a time step $\Delta t$ and a network step $\Delta x = \Delta y$,, we get:

$$\partial\varphi(x,y,t)/\partial t = (D/4)\nabla^2 \{\theta[\varphi(x+1,y,t) - \varphi_s]\} + \eta(x,y,t),$$

where $\nabla^2$ is the Laplacian operator, $D = (\Delta x)^2/\Delta t$ the diffusion coefficient and $\eta(x,y,t)$ is the additionnal adiabatic noise.
The difficulty in this approach of the threshold dynamics appear when developping the spatial derivatives. We obtain:

$$\partial\varphi/\partial t = (D/4)\left\{\delta[\varphi - \varphi_s]\nabla^2\varphi + \Delta t^{-1}\delta'[\varphi - \varphi_s](\nabla\varphi)^2\right\} + \eta(x,y,t),$$

where $\delta[\varphi - \varphi_s]$ and $\delta'[\varphi - \varphi_s]$ are respectively the Dirac function and its first derivative. The most important aspect of this equation is the existence of a Dirac peak of the diffusion coefficient. Then a non-linear term $(\nabla\varphi)^2$ multiplied by a divergent amplitude when $\varphi = \varphi_s$ produce instabilities.
The accelerate dynamics in the instability threshold neighbourhood is the cause of attraction of any initial condition to a self organized critical point.

# Chapter 12

# Multifractals

## 12.1 Generalized Dimensions of Fractals and Strange Attractors

Following Hentschel and Procaccia (1983) [99], let us define the three best known dimensions: the similarity dimension $D$, the information dimension $\sigma$ and the correlation dimension $\nu$ (see also [75] and chapter 9).
We consider an attractor which is embedded in a $d$-dimensional space. Let $\{X_i\}_{i=1}^N$ be the points of a long time series, where $N$ is very large but necessarily finite, on the attractor. When covering phase space with a mesh of $d$-dimensional cubes of size $b^d$ we get a number of $M(b)$ cubes that contain the points of the series $\{X_i\}_{i=1}^N$. Let us define $p_k \equiv N_k/N$ where $N_k$ is the number of point in the $k$th cube.
The **similarity dimension** is defined by [128, 75]:

$$D = -\lim_{b\to 0}\lim_{N\to\infty} \log M(b)/\log b\,. \tag{12.1}$$

The similarity dimension is the fractal dimension as it was defined by Mandelbrot [128].
The **information dimension** is defined by [14, 55]

$$\sigma = -\lim_{b\to 0}\lim_{N\to\infty} S(b)/\log b\,. \tag{12.2}$$

where

$$S(b) = -\sum_{k=1}^{M}(b) p_k \log p_k\,. \tag{12.3}$$

The **correlation dimension** is defined by [75, 127, 173]

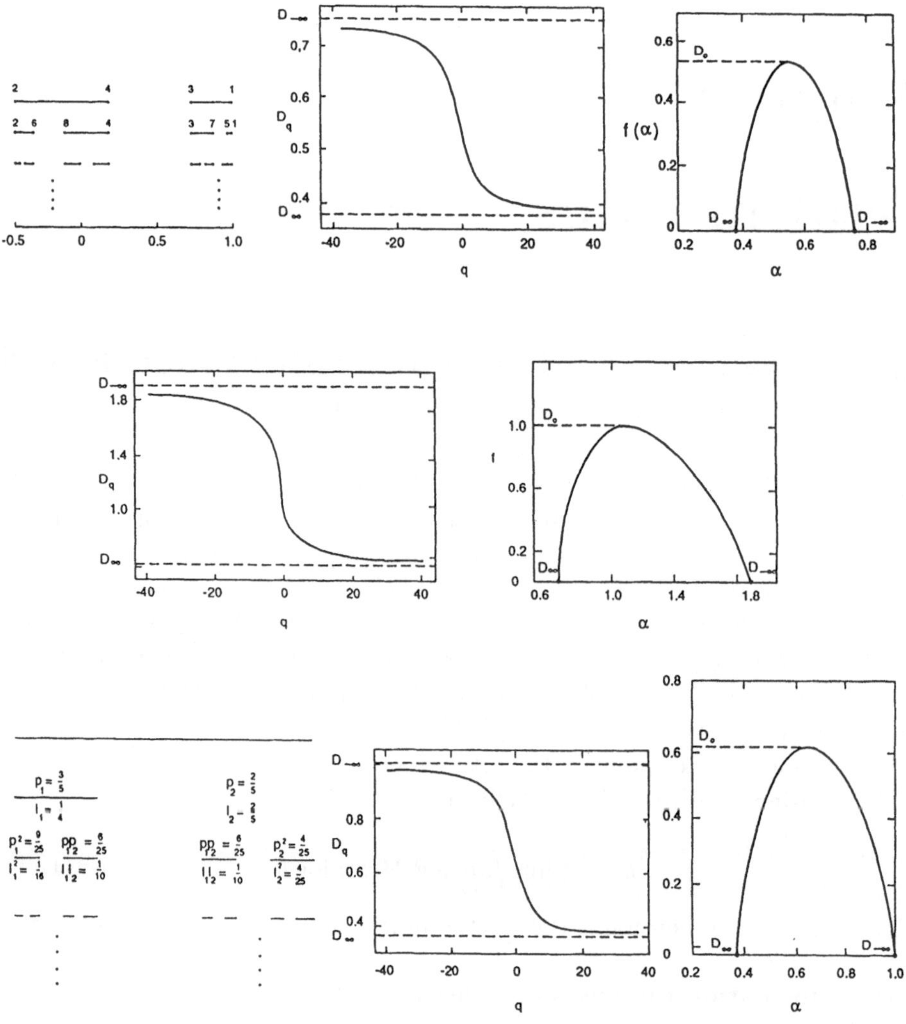

Figure 12.1: *Generalized dimensions.* The (after Halsey *et al.*, 1986).

$$\nu = -\lim_{b\to 0}\lim_{N\to\infty} \log C(b)/\log b\,, \tag{12.4}$$

where

$$C(b) = \frac{1}{N^2}\sum_{i\neq j}\theta(b- \mid \mathbf{X}_i - \mathbf{X}_j \mid)\,. \tag{12.5}$$

$\theta$ is the Heaviside function (see chapter 9) and $C(b)$ is the correlation integral which classify pairs of points whose distance $\mid \mathbf{X}_i - \mathbf{X}_j \mid$ is less than $b$.

Hentschel and Procaccia (1983) [99] have shown that the definition can be generalized and that there exists a hierarchy of **generalized dimensions** $D_q$ which are defined for any $q \geq 0$.

The general formulation of generalized dimension or **Renyi dimension** is given as follows :

$$D_q = \frac{1}{q-1}\lim_{b\to 0}\frac{\log \sum_{i\in q} p_i^q}{\log b}\,, \tag{12.6}$$

where $p_i$ is the probability to find a point in the box $b$.

Furthermore, the authors [99] have shown that if

$$\lim_{q\to 0} = D\,,$$

then

$$\lim_{q\to 1} = \sigma\,,$$

then

$$D_{q=2} = \nu\,,$$

and for $q = 3,\ 4, \cdots,\ n$ we obtain generalized dimensions associated with correlation integrals of triplets, quadruplets ... and $n$-tuplets of points of the attractor.

It was demonstrated that:

$$D_q > D_q' \quad \text{for any} \quad q' > q\,.$$

Finally it is possible to generalize the study with values of $q$, not necessarily an integer and varying from $-\infty$ to $+\infty$. Figure 12.1 represents some applications of generalized dimensions for Cantor sets [95, 96].

## 12.2 Definitions and Properties

### 12.2.1 A Simple Definition

Let us take $N$ to be the number of boxes of size $\Delta$ necessary to cover a given set.
When the set is fractal we can write: $N \sim \Delta^{-D_0}$ when$\Delta \to 0$. $D_0$ is the fractal set dimension.
Let $p_i(\Delta)$ be the probability of visiting a size-$\Delta$ box in the vicinity of a set point $x_i$.
For a simple fractal set, we have $p_i(\Delta) = \Delta^{D_0}$ for any $x_i$.
For more complex fractal set we have $p_i(\Delta) = \Delta^{\alpha(x_i)}$ where $\alpha(x_i)$ varies from one point to another point.

$\alpha(x_i)$ is called the *singularity index*, or Holder exponent, or punctual dimension of the fractal set

$$\alpha_i \equiv \alpha(x_i) = \log p_i(\Delta)/\log\Delta, \quad \Delta \to 0\,,$$

where $p_i$ is the non-uniform probability (or measure) of the $x_i$.
The *singularity spectrum* defines a *Multifractal set*
The number of boxes whose exponent is $\alpha(x_i)$ inside the interval $\alpha$, $\alpha+\delta\alpha$ is:

$$N_\Delta(\alpha)\delta\alpha = \rho(\alpha,\Delta)\Delta^{-f(\alpha)\delta\alpha}, \quad \Delta \to 0\,. \tag{12.7}$$

The density $\rho(\alpha,\Delta)$ is a function of $\delta\alpha$.
The function $f(\alpha)$ describes the singularity index spectrum of the multifractal set.
This spectrum can be obtained as well from the generalized dimension for mula

$$D(q) = \lim_{\Delta\to 0} \frac{\log \sum_i p_i^q(\Delta)}{(q-1)\log\Delta}$$

Frisch and Parisi have shown [62] that:

$$\begin{cases} f(\alpha_q) & = q\alpha_q - (q-1)D_{(q)} \\ \alpha_q & = \dfrac{d}{dq}[(q-1)D_{(q)}] \end{cases} \tag{12.8}$$

We recognize here that $f(\alpha)$ and $(q-1)D_{(q)}$ are, respectively, Legendre transforms of each other.
Figure 12.1 shows the singularity spectrum or multifractal spectrum obtained from this relationship

### 12.2.2 A Measure on the Middle Third Cantor Set

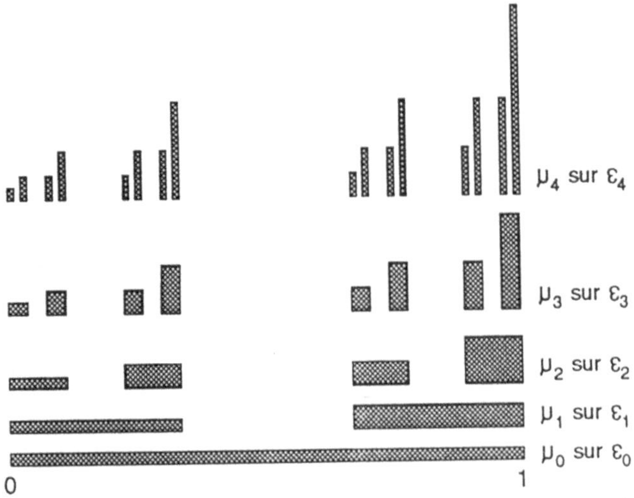


Figure 12.2: *Self similar measure on the middle third Cantor set.* The (after Falconer, 1990).

In this academic example (see Falconer, 1990, [54]) the mass (measure) of each interval on $\mathcal{E}_k$ indicated by the area of the rectangle, is divided in the ratio $1/3; 2/3$ between the two subintervals of $\mathcal{E}_{k+1}$. Continuing this process yields a mass distribution $\mu$ of the Cantor set. On figure 12.2 we can see that the sum of all the elements on each line of iteration is 1. (in terms of probabilities, this sum corresponds to that of all the partial probabilities which is the total probability and is always 1).

The construction of the multifractal spectrum is possible if the distribution of different elements is binomial. Falconer [54] obtained the exact solution of $f(\alpha)$ which is presented in figure 12.3

### 12.2.3 Practical Methods of Constructing a Multifractal Spectrum

The goal of the methods is to construct the diagram of $f(\alpha)$ *vs* $\alpha$. There are two main methods:

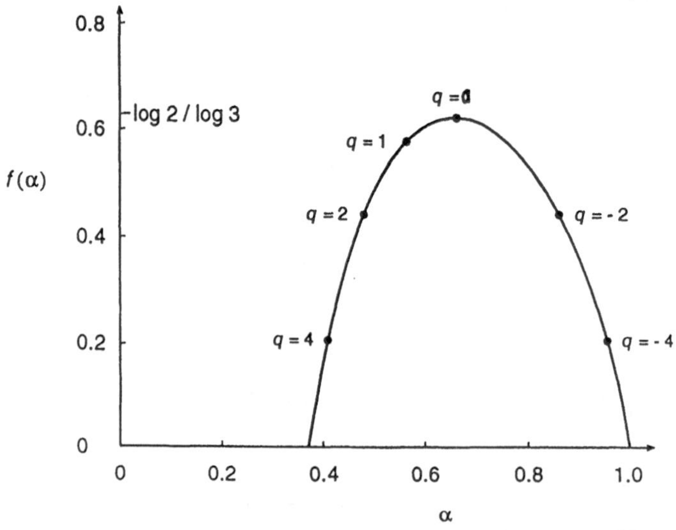


Figure 12.3: *Singularities spectrum of a the middle third Cantor support multifractal set.* In this example the support is the middle third Cantor set and the measure is defined on the previous figure (after Falconer, 1990).

## The Histogram Method

Here we enumerate the successive steps of the method.
Given a measure $\mu$ on a fractal set (the support)

1. we begin to distribute the measure in boxes of size $\varepsilon$ and that gives us a collection of boxes
$$\{\mathcal{B}_i(\varepsilon)\}_{i=1}^{N(\varepsilon)}$$
where $N(\varepsilon)$ is the number ob boxes necessary to cover the entire support of the measure.

2. if $\mu(\mathcal{B}_i)$ is the box $i$ measure, then we can compute the Holder exponent
$$\alpha_i = \log \mu_i / \log \varepsilon\,.$$

3. to draw the histogram we divide the $\alpha$ axis into segments whose size is $\Delta\alpha$ and we estimate the number $N_\varepsilon(\alpha)$ by plotting the number $N_\varepsilon(\alpha)\Delta\alpha$ of occurence of a specific value between $\alpha$ and $\alpha + \Delta\alpha$,

4. we repeat the process for different values of the box size $\varepsilon$,

5. then, we plot the graph

$$-\log N_\varepsilon(\alpha)/\log\varepsilon \quad vs \quad \alpha$$

for different values of $\varepsilon$, which gives

$$N_\varepsilon(\alpha) \sim \varepsilon^{-f(\alpha)}.$$

When these points are on a curve $f(\alpha)$ for small values of $\varepsilon$, we say that the mesure is multifractal

## The Moments Method

The method is based on the quantity which is called *the repartition function* and defined as:

$$\chi_q(\epsilon) = \sum_{i=1}^{N(\epsilon)} \mu_i^q, q \in \Re. \tag{12.9}$$

In the case of a binomial measure, the repartition function is $\chi_q(\epsilon_k)$ corresponding to the box size $\epsilon_k = 2^{-k}$. We get

$$\begin{aligned}
\chi_q(\epsilon_0) &= 1^q, \\
\chi_q(\epsilon_1) &= m_0^q + m_1^q, \\
\chi_q(\epsilon_2) &= (m_0^q + m_1^q)^2, \\
\text{and more generally } \chi_q(\epsilon_k) &= (m_0^q + m_1^q)^k.
\end{aligned}$$

(see figure 12.4)
Coming back to the previous definition, let us write $\mu_i$ box measures in the form $\mu_i = \epsilon^{\alpha_i}$, which gives

$$\chi_q(\epsilon) = \sum_{i=1}^{N(\epsilon)} (\epsilon^{\alpha_i})^q. \tag{12.10}$$

So there are $N_\epsilon(\alpha)d\alpha$ boxes, out of the total $N(\epsilon)$, for which the Hölder coarse exponent satisfies $\alpha < \alpha_i < \alpha + d\alpha$. Let us assume that there exist constants $\alpha_{\min}$ and $\alpha_{\max}$ such that $0 < \alpha_{\min} < \alpha < \alpha_{max} < \infty$ and that $N_\epsilon(\alpha)$ is continuous. Thus, the contribution to $\chi_q(\epsilon)$ of boxes subset with $\alpha_i$ between $\alpha$ and $\alpha + d\alpha$ is $N_\epsilon(\alpha)(\epsilon^\alpha)^q d\alpha$.
For $\epsilon$ given, we then integrate over $\alpha$

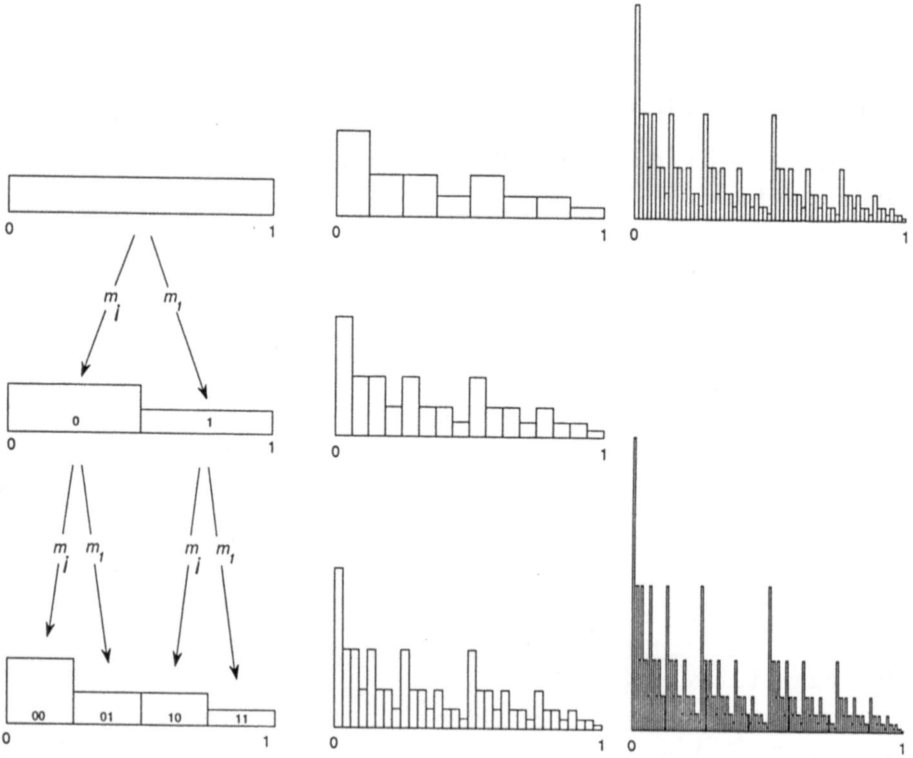


Figure 12.4: *An example of cascade that generates a binomial measure.* We have chosen a dyadic set into which the measure is such that : $m_1 = 1 - m_0$ and $m_0 = 2/3$ ; $m_1 = 1/3$ . The cascade was iterated to order 8 (after Evertsz and Mandelbrot, 1990).

$$\chi_q(\epsilon) = \int N_\epsilon(\alpha)(\epsilon^\alpha)^q d\alpha\,. \tag{12.11}$$

When $N_\epsilon(\alpha) \sim \epsilon^{-f(\alpha)}$, we get

$$\chi_q(\epsilon) = \int \epsilon^{q\alpha - f(\alpha)} d\alpha\,. \tag{12.12}$$

To the limit, when $\epsilon \to 0$, the prevailing contribution of the integral comes from $\alpha$ close to the value which minimizes $q\alpha - f(\alpha)$. If $f(\alpha)$ is differentiable, the nécessary condition to have an extrmum is

$$\frac{\partial}{\partial\alpha}\{q\alpha - f(\alpha)\} = 0\,. \tag{12.13}$$

For a given value of $q$, we get that extremum for $\alpha = \alpha(q)$ such as

$$\frac{\partial}{\partial\alpha} f(\alpha) \mid_{\alpha=\alpha(q)} = q\,, \tag{12.14}$$

and this extremum is minimum if

$$\frac{\partial^2}{\partial\alpha^2} f(\alpha) \mid_{\alpha=\alpha(q)} < 0\,. \tag{12.15}$$

Thus, (fig. 12.5) the function $f(\alpha)$ must be convex and for $\alpha = \alpha(q)$, where the minimum is attained, the slope of $f(\alpha)$ is $q$.
If we come back to the contribution dominant in the equation we define

$$\tau(q) = q\alpha(q) - f(\alpha(q))\,, \tag{12.16}$$

which gives

$$\chi_q(\epsilon) = \epsilon^{\tau(q)}\,. \tag{12.17}$$

In the binomial measures application we obtain

$$\tau(q) = \lim_{k\to\infty} \frac{\log(m_0^q + m_1^q)^k}{\log 2^{-k}} = -\log_2(m_0^q + m_1^q)\,, \tag{12.18}$$

and, for the multinomial measures

$$\tau(q) = \log_b \sum_{i=0}^{b-1} m_i^q\,. \tag{12.19}$$

When recalling equation 12.16, we see that

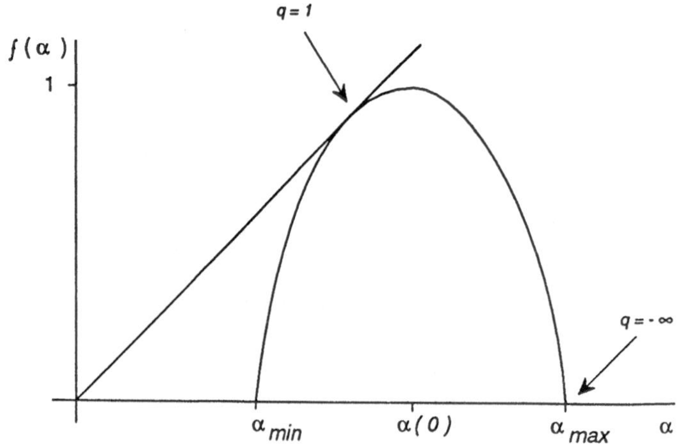


Figure 12.5: *Multifractal spectrum of the binomial measure.* The graph $f(\alpha)$ of the binomial measure defined on the previous figure. We have plotted the point where $q = 1$ (Evertsz and Mandelbrot, 1990).

$$\frac{\partial}{\partial q}\tau(q) = \alpha(q)\,. \tag{12.20}$$

Furthermore, we see that $f(\alpha)$ may be obtained from $\tau(q)$, and vice versa, due to the identity

$$f(\alpha(q)) = q\alpha(q) - \tau(q)\,. \tag{12.21}$$

This relationship between $\tau(q)$ and $f(\alpha)$ is a *Legendre transform.*
We also remark that the function $\tau$ can be related to the expression of generalized dimension $D_q$, which we defined in section 8.7, by $\tau(q) = (q-1)D_q$.
Practically, to compute $f(\alpha)$ by the moments method, we proceed in the following way:

1. We cover coarsely the measure by $\epsilon$ sized boxes, such as $\{B_i(\epsilon)\}_{i=1}^{N}(\epsilon)$ and we determine the measures by box $\mu_i = \mu(B_i(\epsilon))$.

2. We calculate the repartition function, from the equation 12.9, for different values of $\epsilon$.

3. We examine if $\log \chi_q(\epsilon)$ is a linear function. If so, $\tau(q)$ is the line slope corresponding to $q$ exponent(equation 12.16).

4. We calculate $f(\alpha)$ the Legendre transform of $\tau(q)$ (equation 12.21).

Practically, these successive steps are performed numerically, because it is not possible to compute analytical expressions an exception being the triadic multifractal set solved by Falconer (1990)).

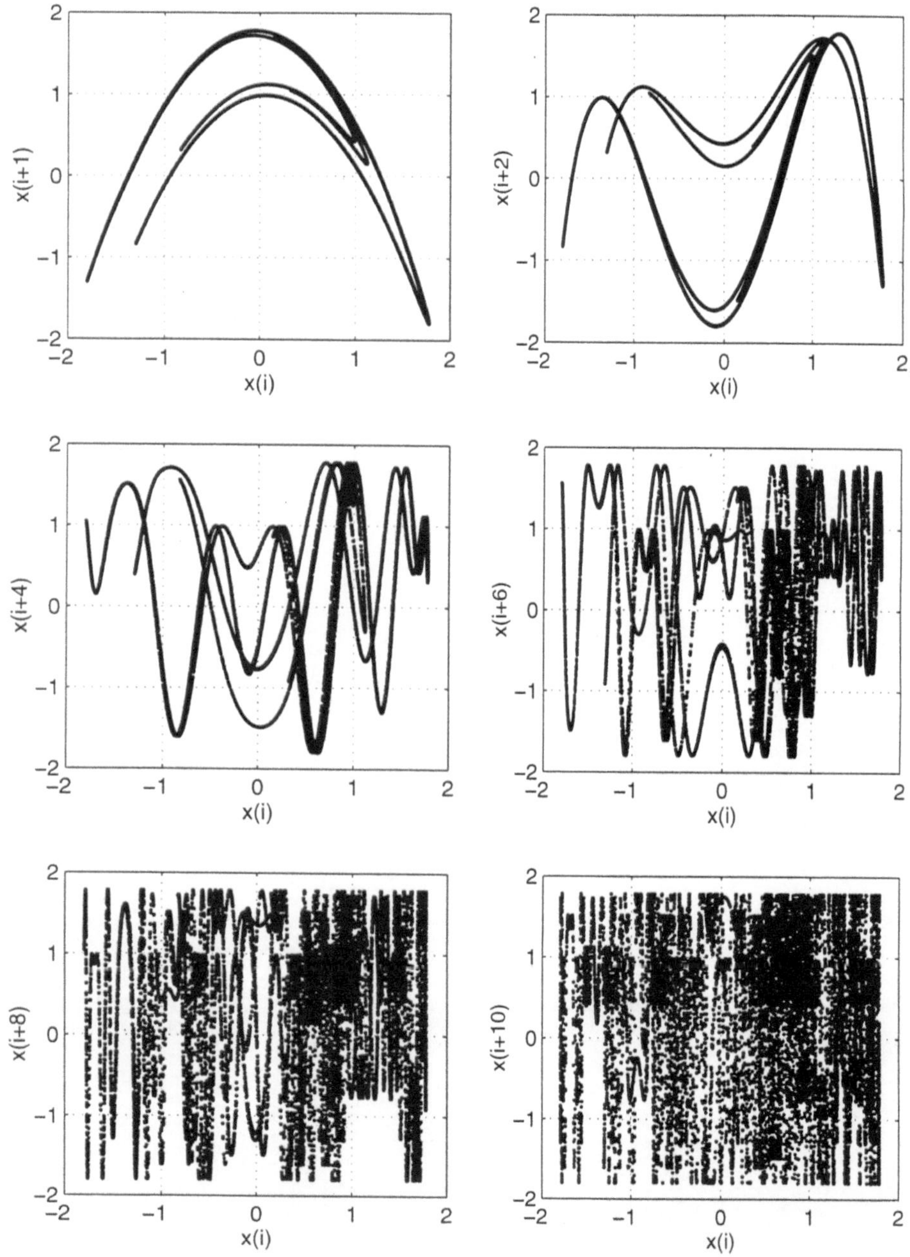


Hénon mapping - The first return mapping is presented in the first graph. On the following graphs we see the increasing non-correlation processes when considering return mappings of order 2, 4, 6, 8, and 10.(series of 20 000 points, after David Aubert, 1997).

# Part IV

# Convex Programming and Systems of Rigid Blocks with Deformable Layers

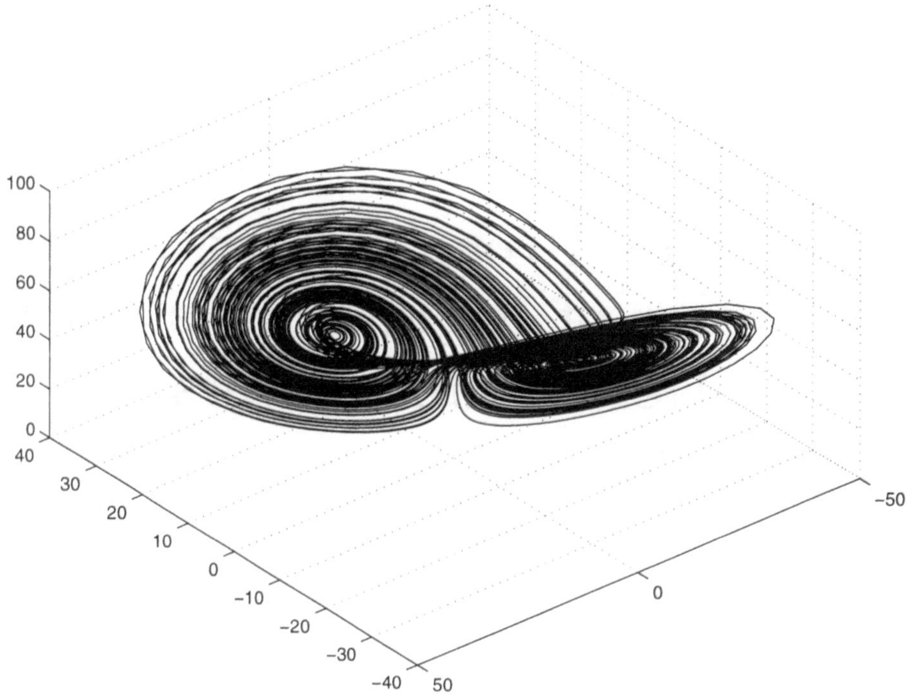


The Lorenz Attractor – The parameters are the same as in page 4, but the perspective representation is different. It shows in the phase space shows 10,000 points discrete phase trajectories (after David AUBERT, 1997).

# Chapter 13

# Systems of Rigid Blocks with Thin Deformable Layers (SRBTDL)

We consider a system of absolutely rigid bodies (rigid blocks) and thin deformable layers which distinguish the neighboring blocks. Such a system allows us to model (in many cases) the processes that characterize the lithosphere dynamics, rock mechanics and some important engineering constructions such as dams, mines, buildings etc.

The idea of dynamic and quasi-static process modeling in non-homogenious media by considering absolutely rigid fragments (rigid blocks) was used already in 1971 by P. Cundall in [38] to analyse large scale movements in a blocky rock sýstem. Later on, in [64], such an approach was used by Gabrielov *et al.* to model seismic processes.

A new point in our approach is application to the quasi-static case of the well-developed convex programming techniques [83].

Before giving a formal definition of System of Rigid Blocks with Thin Deformable Layers (SRBTDL) we illustrate it here by possible applications. For such illustrations we need the following information about SRBTDL.

We assume that blocks $V = \{v_1, \cdots, v_n\}$ are absolutely rigid and layers $E = \{e_1, \cdots, e_m\}$ are absolutely elastic. The last means that the deformation energy of SRBTDL is defined by its current state and does not depend on the history. At the same time the layers can be not homogeneous and "non-linear". In other words, elastic properties can change from one point to another and the Hooke law must hold. Furthermore, most of the results given below are true only in the case of small deformations of SRBTDL.

The space of states of SRBTDL is infinite dimensional. Indeed, because the

layers are absolutely elastic, a state of SRBTDL is defined by the blocks positions. On the other hand, a position of a rigid block in $d$-dimensional space is defined by the finite number of parameters $k(d)$. Thus,in practically real cases $k(d) = 1,\ 3,\ 6; k(1) = 1,\ k(2) = 3,\ k(3) = 6$.
In principle, the space of deformations even of one layer has infinite dimension. However, in SRBTDL model only a limited part of the deformations which correspond to the movement of rigid blocks is considerd.
To each block $v \in V$ and layer $e \in E$ we correspond two vectors
$\mathbf{v} \to (x_v,\ x_v^\star), \quad \mathbf{e} \to (y_e,\ y_e^\star)$ where :

- $x_v$-the coordinates of the block $v$ in the space $\Re^d$,
- $x_v^\star$-total external force applied to $v$,
- $y_e$-deformation of the layer $e$,
- $y_e^\star$-total force of elastic interaction of the layer $e$ with neighbouring blocks.

Furthermore, we introduce $n$-dimensional vectors and $m$-dimensional vectors :

$$\begin{aligned} \mathbf{x} &= (x_{v_1},\ \cdots,\ x_{v_n}), & \mathbf{x}^\star &= (x_{v_1}^\star,\ \cdots,\ x_{v_n}^\star) \\ \mathbf{y} &= (y_{e_1},\ \cdots,\ y_{e_m}), & \mathbf{y}^\star &= (y_{e_1}^\star,\ \cdots,\ y_{e_m}^\star) \end{aligned} \tag{13.1}$$

The vectors (13.1) characterize correspondingly

- $x$-state of SRBTDL,
- $x^\star$-external forces,
- $y$-deformations,
- $y^\star$-internal elastic forces.

The definitions are based on the following inclusions:

$$x,\ x_v^\star, \quad y_e,\ y_e^\star \in \Re^k; \quad x,\ x^\star \in \Re^k n; \quad y,\ y^\star \in \Re^k m\,,$$

where $k = k(d) = d(d+1)/2$. In particular, $k = 1,\ 3,\ 6$, in one, two and three dimensional cases correspondingly.
There exists a matrix $\mathbf{A}$ such that $\mathbf{y} = \mathbf{A}\mathbf{x}$, and $\mathbf{x}^\star = \mathbf{A}^\star\mathbf{y}^\star$, where $\mathbf{A}^\star$ is the transpose of $\mathbf{A}$. The first equation, in some cases, takes place approximately

only when the deformations $y$ are small. The second equation is a precise one. It is the condition of the blocks equilibrium.
The main functions and their subdifferentials are interpreted in the following way :

- $\mathcal{F}_e(y_e)$-elastic energy of the layer $e$ as a function of its deformation $y_e$,
- $y_e^\star = f_e(y_e) = \partial\mathcal{F}_e(y_e)$-global elastic characteristic of the layer $e$,
- $\mathcal{G}_v(x_v)$-potential energy of the block $v$ in the field of external forces as a function of its position,
- $x_v^\star = g_v(x_v) = \partial\mathcal{G}_v(x_v)$-dependence of global external forces on the coordinates of the block $v$,
- $\mathcal{F}(y) = \sum_{e\in E}\mathcal{F}_e(y_e)$-complete internal elastic energy of SRBTDL as a function of deformation $y$,
- $\mathcal{G}(x)$-complete energy of SRBTDL in the field of external forces as a function of its position $x$,
- $\mathcal{L}(x) = -\mathcal{G}(x) + \mathcal{F}(\mathbf{A}x)$-complete potential energy of SRBTDL as a function of its position $x$ (Lagrange function).

The calculation of the static equilibrium of SRBTDL is a problem of convex programing.

Herein, we list possible applications of SRBTDL.

1. **Lithospheric dynamics**. SRBTDL allows to model tectonic processes, if we consider $V$ as lithospheric blocks of different ranks and $E$ as fault zones.
2. **Rock mechanics**. Here we make the scale smaller and apply SRBTDL to problems of the mechanics of the rock failure, in particular for calculations of the rigidity of dams and mines.
3. **Building engineering**. SRBTDL allows to model "block buildings". Here $V$ are large homogeneous parts of the building construction and $E$ are corresponding structural units.

4. **Prediction of failure and calculation of strength**. Induced seismicity in dams that takes place while water fills reservoirs, induced seismicity born from gas deposit exploitation, failure of building constructions as a result of seismicity or other external influences - all these phenomena correspond to failure of deformable layers $E$ as a result of an external load in the SRBTDL model.

Because we assume that blocks $V$ are absolutely rigid, then deformations, stress and elastic energy are fully concentrated in the layer $E$. The techniques proposed here allow us to obtain their distributions in space and time. This gives an opportunuity to predict the place and time of coming destruction and to calculate the strength of SRBTDL.

However, to realize this program, we have to know elastic properties and thresholds of the layers strength in all points, as well as external conditions. Since we do not have this information, the model can give only qualitative picture of the SRBTDL behaviour.

To clarify the role of SRBTDL, we have to distinguish clearly quasi-static and dynamic problems.

Quasi-static processes we have for tectonics (because masses and moments of inertium of the lithospheric blocks are very large) and for rock mechanics in cases, when external factors change slowly; for example, when a water reservoir is slowly filled by water.

Another example is exploitation of gas, oil and mineral deposits. Settling of the soil under the weight of a heavy construction is also a quasi-static process.

Dynamic processes, of course, can also be modelled. However, the methods of convex programming which we propose here can be used only in the static or quasi-static cases.

SRBTDL is described by a system of differential equations. In the general case, a solution of these equations consists of two stages. On the first stage, we consider only static and quasi-static external factors. The state of stable SRBTDL equilibrium is calculated by convex programming technique. On the second stage we include dynamic factors and calculate small SRBTDL oscillations in a vicinity of equilibrium.

Elastic characteristics of layers $y_e^\star = \partial\mathcal{F}_e(y_e), \quad e \in E$ have to be monotonous, but not necessarily linear. Non-linearity gives the possibility to include into consideration the media that "resists compression more than extension" (fig.13.1, a), "compressed in limits" (fig.13.1, b), etc. (see fig.13.1).

Furthermore, we recall that elastic properties of the layers may change from one point to another.

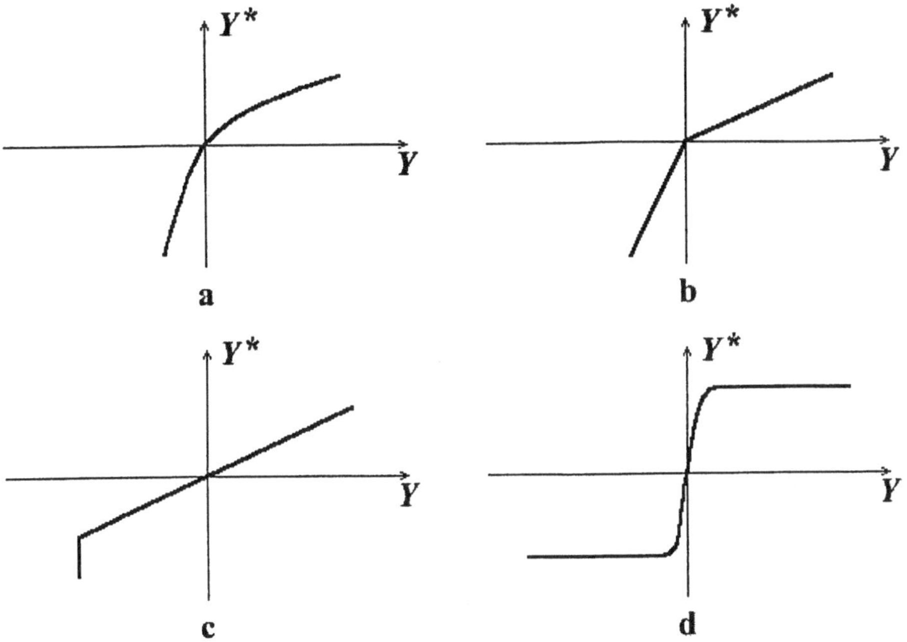


Figure 13.1: *SRBTDL equilibrium.* Non-linearity gives the possibility to include into consideration the media that a) " resists compression more than extension", b) " compressed in limits", etc see text.

External conditions $\widetilde{x_v^\star} = \partial\mathcal{G}_v(x_v)$ are typical for static and quasi-static problems. "Frozen blocks" have fixed positions $x_v = x_v^0$, $v \in V_x$. "Free blocks" $v \in V_x$ are in the field of constant (which do not depend on $x$) external forces $x_v^\star = x_v^{\star 0}$, $v \in V_{x^\star}$. Weight is an example of such kind of external forces.
An example of an external force which depends on the block position $x_v$ is the bouyancy force which is applied to lithosphere blocks sunk into the Earth's mantle.
In the quasi-static case $x_v^0$ and $x_v^{\star 0}$ may depend on time.
The possibility of applying convex programming techniques to an SRBTDL model is based on the following assumptions :

- a) Absolute rigidity of the blocks $v \in V$,
- b) Absolute elasticity of layers $e \in E$,
- c) Convexity of functions $\mathcal{F}_e(y_e)$, $e \in E$ and $\mathcal{G}_v(x_v)$, $v \in V$.

We also assume that the system is sufficiently rigid such that even significant external forces create just small deformations.
Thus, as an input to the problem of calculation of SRBTDL model we have the following information :

- a) Elastic properties of the layers *e.g.* the functions $\mathcal{F}_e$, $e \in E$,
- b) External conditions *e.g.* the functions $\mathcal{G}_v$, $v \in V$,
- c) Geometry of the system *e.g.* the matrices $\mathbf{A}$ and $\mathbf{A}^\star$.

It is clear that in actual applied problems this input information is incomplete or approximate.

# Chapter 14

# System of Rigid and Deformable Blocks (SRDB)

Here we use a more general approach that is not restricted to having only thin deformable layers. Later, we will return to SRBTDL by considering it as a special case of the SRDB.
The main assumptions of the SRDB are the following:

**Mechanics** Rigid blocks $V = \{v_1, \cdots, v_n\}$ are absolutely rigid and deformable blocks $W = \{w_1, \cdots, w_n\}$ are absolutely elastic bodies.
The blocks of the same type ($v \in V$ or $w \in W$) have no common boundaries and do not interact while deformations take place. On the contrary, the different types of blocks may have common boundaries which do interact while deformations take place. In fact, a search for the conditions of the place and time of such connected boundaries is actually our goal. We symbolize by $E = \{e_1, \cdots, e_m\}$ the connected parts of the boundaries.

**Geometry** The blocks under consideration are connected and limited to sets in Euclidean space $\Re^d$.
We do not assume that they are convex and/or uniformly connected.
Figure 14.1 gives an example of SRDB.
As is shown in [], the results described below will remain if Euclidean space $\Re^d$ is replaced by the sphere $S^d$.
The dimension of the space of SRDB states is finite [ ]. A position of a rigid body in $\Re^d$ (or on $S^d$) is defined by $k = d(d+1)/2$ coordinates. Thus, the dimension of the space of SRDB states is equal to $k \cdot n = nd(d+1)/2$. On the line $d = k = 1$ QRDB is practically equivalent to a system of interacting partitions [ ].

We consider the cases $d = 2$ and $d = 3$. They correpond to $k = 3$ and $k = 6$.
Vectors $(x_v, x_v^\star, y_e, y_e^\star)$ which describe SRDB are interpreted in the following way:

- $x_v$-displacement of the block $V$,
- $y_e$-displacement of a sector of the boundary $e$,
- $x_v^\star$-total external force applied to the block $V$,
- $y_e^\star$-total elastic force applied through the boundary $e$ to the neighbouring rigid block.

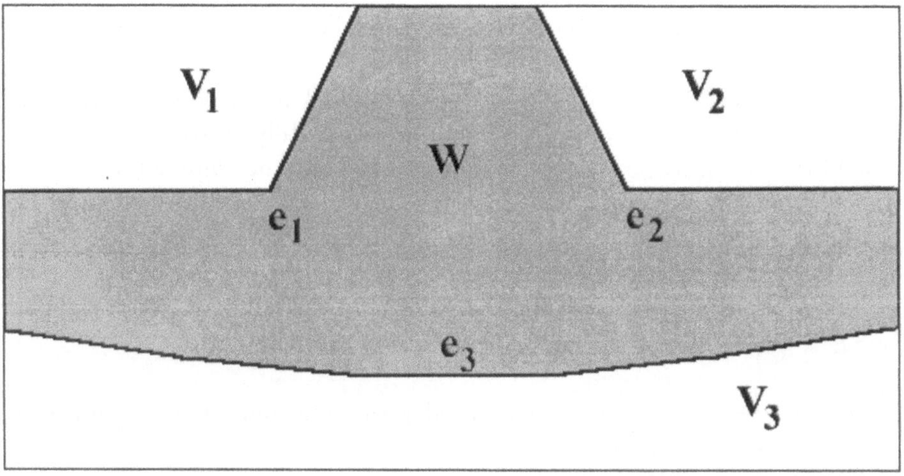


Figure 14.1: *Geometry of three rigid blocks in SRDB.* The three blocks influence each other.

It is clear that $x_v$, $y_e$, $-x_v^\star$, $y_e^\star$ are vectors from $\Re^k$, where $k = k(d)$. Now, we will see what the coordinates of these vectors are.
Any displacement $x_v$ of a rigid body $V$ on the plane $(d = 2)$ or in a three dimensional space $(d = 3)$ is represented in a unique way as a combination of a parallel translation $x_{vT}$ and a rotation $x_{vR}$ around the given centre O.

In this decomposition $x_{vT}$ depends on the choice of O. On the contrary $x_{vR}$ does not.

On the plane $(d = 2)$ circulation is defined by a number which is the angle of the turn. The axis of the rotation in this case is orthogonal to the plane.

In three dimensional space $(d = 3)$ rotation is defined by three dimensional vector, which has the direction equal to the rotation axis and the value equal to the angle of the turn.

Thus, for $d = 2$

$$\begin{array}{ll} x_v = (x_{v1},\ x_{v2},\ x_{v3}) & = (x_{vT},\ x_{vR}) \\ \text{where } x_{vT} = (x_{v1},\ x_{v2}), & x_{vR} = x_{v3} \end{array} \tag{14.1}$$

For $d = 3$

$$\begin{array}{ll} x_v = (x_{v1},\ \cdots,\ x_{v6}) & = (x_{vT},\ x_{vR}) \\ \text{where } x_{vT} = (x_{v1},\ x_{v2},\ x_{v3}), & x_{vR} = (x_{v4},\ x_{v5},\ x_{v6}) \end{array} \tag{14.2}$$

Then we have that $y_e = x_v$, if $e$ is a part of a boundary of the block $v$.

The field of forces applied to a rigid body is characterized by the resulting force and the resulting moment of the forces with respect to the given centre O.

Therefore, for $d = 2$ and $d = 3$ we have correspondingly

$$\begin{array}{ll} -x_v^\star = -(x_{v1}^\star,\ x_{v2}^\star,\ x_{v3}^\star) & = (x_{vT}^\star,\ x_{vR}^\star) \\ \text{where } x_{vT}^\star = (x_{v1}^\star,\ x_{v2}^\star), & x_{vR}^\star = x_{v3}^\star \\ y_e^\star = (y_{e1}^\star,\ y_{e2}^\star,\ y_{e3}^\star) & = (y_{eT}^\star,\ y_{eR}^\star), \\ y_{eT}^\star = (y_{e1}^\star,\ y_{e2}^\star)\ ; & y_{eR}^\star = y_{e3}^\star\ ; \end{array} \tag{14.3}$$

$$\begin{array}{ll} -x_v^\star = (x_{v1}^\star,\ \cdots,\ x_{v6}^\star) & = -(x_{vT}^\star,\ x_{vR}^\star) \\ \text{where } x_{vT}^\star = (x_{v1}^\star,\ x_{v2}^\star,\ x_{v3}^\star), & x_{vR}^\star = (x_{v4}^\star,\ x_{v5}^\star,\ x_{v6}^\star) \\ y_e^\star = (y_{e1}^\star,\ \cdots,\ y_{e6}^\star) & = (y_{eT}^\star,\ y_{eR}^\star), \\ \text{where } y_{eT}^\star = (y_{e1}^\star,\ y_{e2}^\star,\ y_{e3}^\star)\ ; & y_{eR}^\star = (y_{e4}^\star,\ y_{e5}^\star,\ y_{e6}^\star) \end{array} \tag{14.4}$$

Here, $x_{vT}^\star$ and $y_{eT}^\star$ are the forces and $x_{vR}^\star$ and $y_{eR}^\star$ are the moments of the forces,

$$\begin{array}{ll} x \in \Re^k n, & x = (x_{v1},\ \cdots,\ x_{vn}), \\ x^\star \in \Re^k n, & x^\star = (x_{v1}^\star,\ \cdots,\ x_{vn}^\star), \\ y \in \Re^k m, & y = (y_{e1},\ \cdots,\ y_{em}), \\ y^\star \in \Re^k m, & y^\star = (y_{e1}^\star,\ \cdots,\ y_{em}^\star)\,. \end{array} \tag{14.5}$$

For every pair $(v, e)$ we introduce the function

$$inc(v,e) = \begin{cases} 1, & \text{if } e \text{ is a part of the boundary of } v \\ 0, & \text{if not} \end{cases}$$

Following this, we introduce the matrix of incendation

$$\mathbf{A}_{VE} = \| \ inc(v,e) \ \|, \ v \in V = \{v_1, \ \cdots, \ v_n\}, \ e \in E = \{e_1, \ \cdots, \ e_m\}\,.$$

The following linear equations hold

$$\begin{array}{ll} y = \mathbf{A}_{VEK}x & x^\star = \mathbf{A}^\star_{VEK}y^\star \\ \text{where } \mathbf{A}_{VEK} = \mathbf{A}_{VE} \bigotimes E_K, & \mathbf{A}^\star_{VEK} = \mathbf{A}^\star_{VE} \bigotimes E_K - \text{tensor products}\,. \end{array} \tag{14.6}$$

The matrices $\mathbf{A}_{VEK}$ $(ord\mathbf{A}_{VEK} = (kn + km))$ and $\mathbf{A}^\star_{VEK}$ $(ord\mathbf{A}^\star_{VEK} = (km + kn))$ are obtained from the matrices $\mathbf{A}_{VE}$ $(ord\mathbf{A}_{VE} = n \times m)$ and $\mathbf{A}^\star_{VE}$ $(ord\mathbf{A}^\star_{VE} = m \times n)$ by substitution of each element $inc(v,e)$ to the "block" $inc(v,e) \cdot \mathbf{E}_k$, where

$$\mathbf{E}_k = \begin{pmatrix} 1 & & & \\ & 1 & & 0 \\ & & \ddots & \\ 0 & & & \\ & & & 1 \end{pmatrix}$$

The equivalency $y = \mathbf{A}_{VEK}x$ is actually the definition of the vector of displacement $y$. At the same time the equation $y^\star - \mathbf{A}^\star_{VEK}x^\star$ gives us the condition of statical equilibrium of rigid blocks.
Indeed, $-\mathbf{x}^\star$ and $\mathbf{A}^\star_{VEK}y^\star$ are the vectors of the resulting forces and moments created by the application of external and internal (elastic) forces. At equilibrium,

$$-\mathbf{x}^\star + \mathbf{A}^\star_{VEK}y^\star = 0\,.$$

For every block $v \in V$ and a part of the boundary $e \in E$ we construct a particular orthogonal basis and a centre of rotation. Thus, we obtain the set

$$\mathcal{B}_0 = \{\mathcal{B}_v, \ \mathcal{O}_v, \ v \in V, \quad \mathcal{B}_e, \ \mathcal{O}_e, \ e \in E\}\,.$$

Analogous to (1) and (2) we introduce vectors

$$(x_v, x_v^\star;\ y_e, y_e^\star);\ (x_{vT}, x_{vT}^\star;\ y_{eT}, y_{eT}^\star);\ (x_{vR}, x_{vR}^\star;\ y_{eR}, y_{eR}^\star),$$

by changing the pair $(\mathcal{B}_0, \mathcal{O})$ to the pair $(\mathcal{B}_v, \mathcal{O}_v)$ or $(\mathcal{B}_e, \mathcal{O}_e)$. We introduce the vectors $(x, y; x^\star, y^\star)$ by the formula 14.5. All the dimensions have the same values as above. We have precise coincidence of the definitions if

$$\mathcal{B}_0 = \mathcal{B}_v = \mathcal{B}_e,\ \mathcal{O} = \mathcal{O}_v = \mathcal{O}_e,\ \forall v \in V\ \forall e \in E\,.$$

For $d = 2,\ 3$, the following formulas hold

$$y_{eT} = \sum_v \mathcal{B}_{ve}(x_{vT} + x_{vR} \times \mathcal{O}_{ve}) inc(v,e) \tag{14.7}$$

$$y_{eR} = \sum_v \mathcal{B}_{ve} x_{vR} inc(v,e) \tag{14.8}$$

$$x_{vT}^\star = \sum_e \mathcal{B}_{ve}^\star \mathcal{Y}_e T^\star inc(v,e) \tag{14.9}$$

$$x_{vR}^\star = \sum_e \mathcal{B}_{ve}^\star (y_{eR}^\star + y_{eT}^\star \times \mathcal{O}_v E) inc(v,e) \tag{14.10}$$

Here $\mathcal{O}_{ve}$ is a vector which begins in $\mathcal{O}_v$ and ends in $\mathcal{O}_e$, represented in the bases $\mathcal{B}_e$, $\times$ represents the vector product, and $\mathcal{B}_{vE} - (d \times d)$ is a transfer matrix from the bases $\mathcal{B}_v$ to the bases $\mathcal{B}_e$.
Because all the bases are orthonormal $\mathcal{B}_{vE}$ is an orthogonal matrix:

$$\mathcal{B}_{ve} \cdot \mathcal{B}_{ve}^\star = Ed \text{ or } \mathcal{B}_v E^\star = \mathcal{B}_{ve}^{-1}\,.$$

The formulas 14.9, and 14.10 represent $x_v^\star$ through $y_e^\star$ according to the laws of statics. External forces and their moments applied to a block $v$ as well as the internal elastic forces and their moments give the equilibrium on the boundary.
The formulas 14.7, 14.8 are purely kinematic. They express $y_e$ through $x_v$, taking into account the change of $\mathcal{B}_v$ to $\mathcal{B}_e$ and of $\mathcal{O}_v$ to $\mathcal{O}_e$ while the transfer from $v$ to $e$ takes place. It should be emphasized that the formula 14.9 is approximate, because the displacement of the boundary $e$ while the block $v$ turns (in case $inc(v,e) = 1$) is exactly described by a vector product only when the angle of the turn $\mid x_{vR} \mid$ is small.
From the formulas 14.7 - 14.10, we obtain the following linear equations

$$\begin{aligned} y &\simeq \mathbf{A}_{VEBO}\, x \\ x^\star &= \mathbf{A}_{VEBO}^\star\, y^\star \end{aligned} \tag{14.11}$$

analogous to formulas 14.6. The matrices $\mathbf{A}_{VEBO}$ and $\mathbf{A}^{\star}_{VEBO}$ can be obtained from $\mathbf{A}_{VE}$ and $\mathbf{A}^{\star}_{VE}$ by substitution of each element $inc(v,e)$ to a square $(k \times k)$-matrix $inc(v,e) \cdot \mathbf{A}_{ve}$. The matrices $\mathbf{A}_{ve}$ for $d = 2$, 3 have the following appearence

$$d=2 \qquad \begin{array}{|cc|c|}\hline & & 0 \\ \multicolumn{2}{|c|}{\mathbf{B}_{ve}} & 0 \\ -a_{ve2} & a_{ve1} & 1 \\ \hline \end{array} \quad = \quad \mathbf{A}_{ve}\,,$$

$$d=3 \qquad \begin{array}{|c|c|}\hline \mathbf{B}_{ve} & 0 \\ \hline \mathbf{C}_{ve} \cdot \mathbf{B}_{ve} & \mathbf{B}_{ve} \\ \hline \end{array} = \mathbf{A}_{ve}\,,$$

$$d=3 \qquad \begin{array}{|c|c|c|}\hline 0 & a_{ve3} & -a_{ve2} \\ \hline -a_{ve3} & 0 & a_{ve1} \\ \hline a_{ve2} & -a_{ve1} & 0 \\ \hline \end{array} \quad = \quad \mathbf{C}_{ve}\,.$$

It is clear that $\mathbf{B}^{\star}_{ve} = \mathbf{B}^{-1}_{ve}$, $\mathbf{C}^{\star}_{ve} = -\mathbf{C}_{ve}$.
If a part of the boundary $e$ is a strait segment $(d = 2)$ or a part of a plane $(d = 3)$, then it is convenient to choose a basis $\mathbf{B}_e$ such that its first vector is orthogonal to $e$. In this case we have that for $d = 2$:

- $y_{e1}$, $y^{\star}_{e1}$- are normal displacement and force
- $y_{e2}$, $y^{\star}_{e2}$- tangential displacement and force
- $y_{e3}$, $y^{\star}_{e3}$- turn and moment of forces

For $d = 3$ the following interpretation comes

- $y_{e1}$, $y^{\star}_{e1}$- normal displacement and force
- $(y_{e2}, y_{e3})$- tangential displacement
- $(y^{\star}_{e2}, y^{\star}_{e3})$- tangential force
- $(y_{e4}, y_{e5})$- trend
- $(y^{\star}_{e4}, y^{\star}_{e5})$-"trending" moment of forces

- $(y^{\star}_{e5},\ y^{\star}_{e6})$- circulation and circulating moment of forces.

The energy $F(y)$ of elastic deformation of SRDB is the sum of energies of particular deformable blocks *e.g.*

$$F(y) = \sum_{w} F_w(y_w)\,.$$

But, it is not separable because $F(y) \neq \sum_e F_e(y_e)$. The energy is invariant to movements of the blocks $w \in W$ without deformation.
The apparent formulas and conditions of convexity stability for the function $F(y)$ will be obtained in the next chapter.
The energy of a SRDB has the following important property of additivity:

$$F(y) = \sum_{w \in W} F_w(y_w), \quad F^{\star}(y^{\star}) = \sum_{w \in W} F^{\star}_w(y^{\star}_w)\,. \tag{14.12}$$

Vectors $y_w$ and $y^{\star}_w$, that characterize, correspondingly, deformations and elastic forces applied to blocks $w \in W$, have the appearence:

$$y_w = (y_e \ : \ e \mid inc(w,e) = 1)$$

$$y^{\star}_w = (y^{\star}_e \ : \ e \mid inc(w,e) = 1)$$

Let us consider the situation represented in figure 14.2. Three rigid blocks influence each other, but it cannot be represented as a sum of three bi-lateral influences.
Formally, for the block $V$, the equivalencies

$$F_w(y_w) = \sum_{e \in E} F_e(y_e) inc(w,e),$$

$$F^{\star}_w(y^{\star}_w) = \sum_{e \in E} F^{\star}_e(y^{\star}_e) inc(w,e)$$

may not hold.
Thus, the equivalencies

$$F(y) = \sum_{e \in E} F_e(y_e), \quad F^{\star}(y^{\star}) = \sum_{e \in E} F^{\star}_e(y^{\star}_e)\,. \tag{14.13}$$

that reflect blocked separable property of SRDB also may not take place.

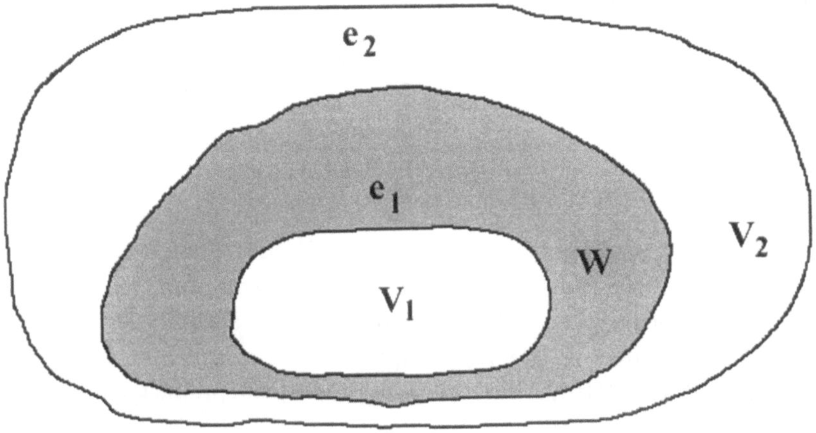


Figure 14.2: *Geometry of three rigid blocks in SRDB.* The three blocks influence each other, but the situation cannot be represented as a sum of three bi-lateral influences.

Therefore, contrary to SRBTDL, an SRDB is not block-separable. We assume that the potential energy of deformable blocks in the field of external forces is negligible. We need this assumption to apply Rockefeller diagrams to SRDB models. In particular, the assumption holds if the masses of deformable blocks are negligible compared to the masses of the rigid blocks. The last is true for SRBTDL. The functions $F_w(y_w)$ have special appearance. They have to be invariant towards transitions of the block $w \in W$ without deformations. Let $y'_w$ and $y_w$" be the positions of the block $w$ before and after its transition. Then, $F_w(y'_w) = F_w(y_w")$. Thus, the energy of elastic deformations is invariant to the movements of the blocks $w \in W$.

Conditions of equilibrium of deformable blocks follow from the facts that the potential energy of deformable blocks in the field of external forces is negligible and the energy is invariant to the movements. Therefore, the equivalencies $y = \mathbf{A}x$ and $y^\star = \mathbf{A}^\star x^\star$ guarantees the equilibrium not only for the rigid blocks, but for all SRDB.

Apparent formulas for elastic energy $F_w(y_w)$ and the conditions of its convexity, unfortunately, are possible to obtain only in some particular cases. The most important among them are the cases of thin linear ($d = 2$) and plane ($d = 3$) layers. Other important cases are when the deformable block $w$ are polygons ($d = 2$) or spatial multifaces figure ($d = 3$), the faces of

which have common boundaries with rigid blocks.
The finite elements method for a static problem of elastic theory and Rockefeller's diagrams for SRDB are closely connected. To show this, we divide a given absolutely elastic body into number of polygons ($d = 3$) or spatial multifaces figures ($d = 3$), which play the role of deformable blocks. We put material points in the apexes and they play the role of rigid blocks. Thus we have again SRDB.
Knowing the elastic properties of a body $S$ and the boundary conditions of the initial problem, we calculate the functions $F(y)$ and $G(x)$ for the SRDB. Then, we search for the positions of the rigid blocks which guarantee the equilibrium. In this way, we obtain an approximate solution of the problem. It is clear that a smaller division of the body $S$ leads to a better approximation.
Contrary to SRBTDL, where rigid blocks occupy most of the system volume, in an SRDB the volume of rigid blocks is very small. It is possible to write linear equations for the energy of elastic deformation of a homogeneous polygon that satisfies Hooke's law. The coordinates of the vectors $y$ and $y^\star$ that correspond to rotation and force moments, are equal to zero because the rigid blocks in this case are the material points.
The SRDB dynamics is analogous to the dynamics of a system of interacting partitions.
The dependance function $\Psi$ expresses Newton's second law for rigid bodies:

$$\Psi(\ddot{x},\ \dot{x},\ x,\ x^\star;\ t) = M\ddot{x} + g(\dot{x},\ x,\ t) - x^\star = 0\,, \tag{14.14}$$

where the diagonal matrix $M$ defines the masses and moments of inertia of rigid blocks. More precisely, if $d = 2$ then the diagonal elements $M[3(i-1)+j;\ 3(i-1)+j]$ expresses the mass of the block $V_i$ for $j = 1,\ 2$ and its moment of inertia for $j = 3$; $M - (3n \times 3n)$-matrix; $i = 1,\ 2,\ 3$. If $d = 3$ the diagonal elements $M[6(i-1)+j,\ 6(i-1)+j]$ expresses the mass of the block $V_i$ for $j = 1,\ 2,\ 3$ and the components of its vector moments of inertia for $j = 4,\ 5,\ 6$; $M = (6n \times 6n) -$ matrix; $i = 1,\ \cdots,\ n$; $j = 1,\ \cdots,\ 6$.
The dependance that defines elastic and plastic properties of deformable blocks formulates a law of rigid-block interaction.
For absolutely elastic thin layers an equivalency $y^\star = \partial F(y)$ takes place. We shall obtain the formulas for elastic energy $F$ in the next paragraph. If external forces have the appearance $x^\star = \partial G(x)$, then the law of minimum action holds.

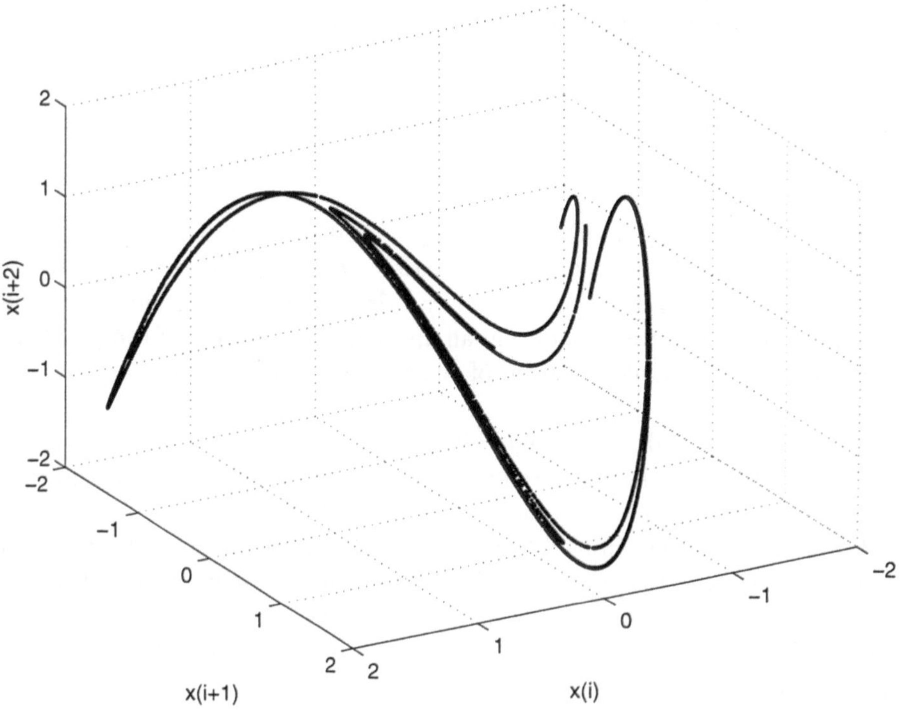


A 3D Hénon mapping : $(x(i),\ x(i+1),\ x(i+2))$ for a 10 000 points series (after David Aubert, 1997).

# Part V

# Bibliography

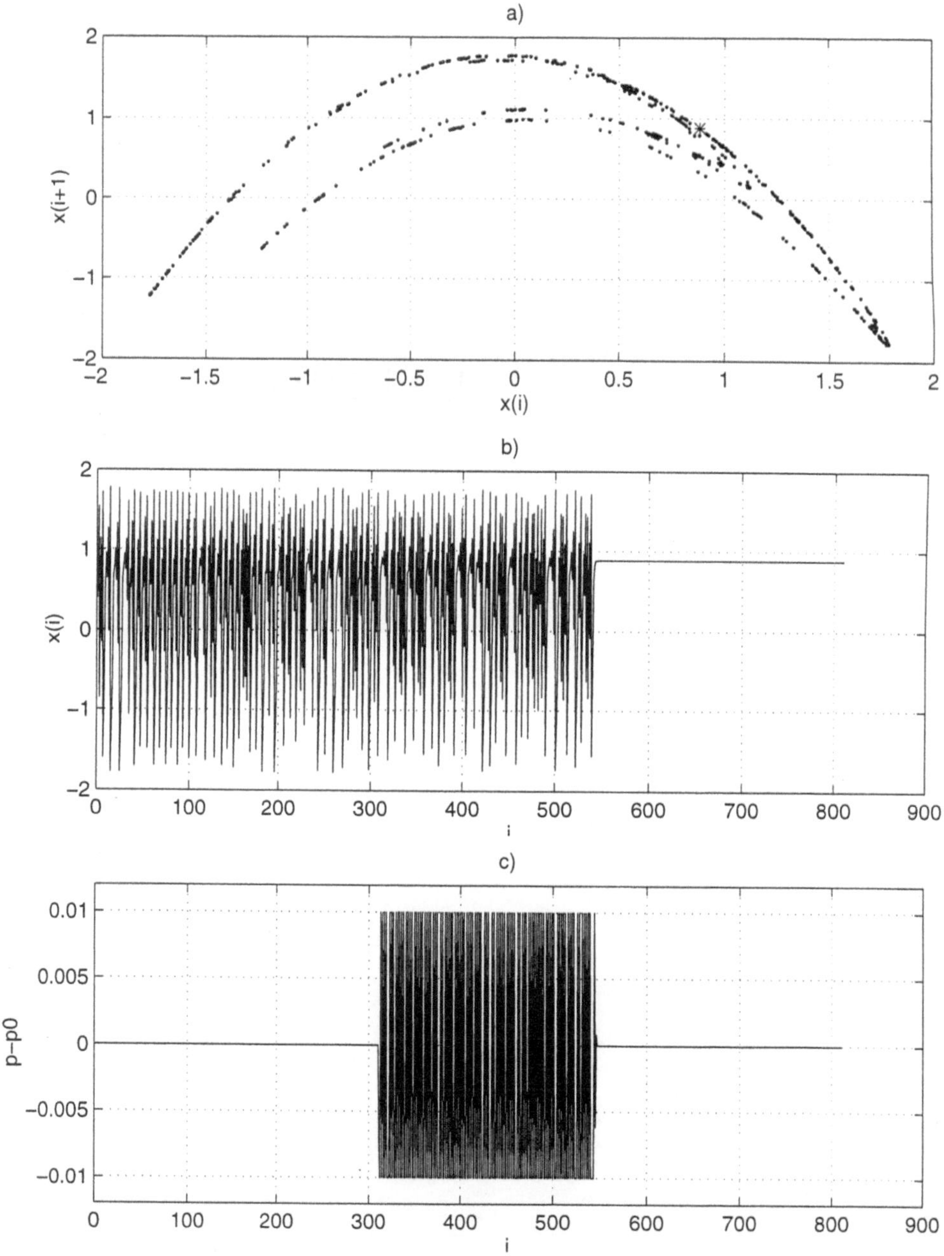


Control of chaos - On a Hénon mapping, we start from $y(i+1) = x(i)$ ; $x(i+1) = p_0 + 0.3y(i) - x(i)^2$ where $p_0 = 1.4$ Then after 550 iterations we changed very slightly $p_0$ (less than 0.01). We observe that the system becomes stabilized (after David Aubert, 1997).

# Bibliography

[1] Ackenhuse, J., 1995. *Signal Processing Technology and Applications: Selected Conference papers*, New York: IEEE Technical Activities Board, 531 p.

[2] Alekseevskaya, M., Gabrielov, A., Gvishiani, A., Gelfand, I. and Ranzman, E., 1977, Formalized Morphostructural Zoning of the Mountain Countries. *Computational Seismology, Moscow, Nauka*, **10**, 227–233.

[3] Allègre, C.J., Le Mouel, J.L. and Provost, A., 1982, Scaling Rules in Rock Fracture and Possible Implications for Earthquake Prediction. *Nature*, **297**, 47–49.

[4] Allègre, C.J., and Le Mouel, J.L., 1994, Introduction of Scaling Techniques in Brittle Fracture of Rocks. *Phys. Earth Plan. Int.* **87**, 85–93

[5] Allègre, C.J., Le Mouel, J.L., Chau, H.D. and Narteau, C., 1995, Scaling Organization of Fracture Tectonics (SOFT) and Earthquake Mechanism. *Phys. Earth Plan. Int.* **92**, 215–233

[6] Arnol'd, V.I., 1992, *Catastrophe Theory*, Springer-Verlag. 150 p.

[7] Artemiev, M., Rotwain, I. and Sadovsky, A., 1977, Recognition of Places where Strong Earthquake May Occur VII. Consideration of Bouguer Gravity Anomalies for California and Adjacent Regions. *Computational Seismology, Moscow, Nauka* **10**. 19–33.

[8] Atten, P. et Caputo, J.G., 1987. Estimation expérimentale de dimension d'attracteurs et d'entropie, in, *Traitement théorique des attracteurs étranges, éd. Cosnard, M., Éditions du CNRS, Paris*, 177–191.

[9] Bak, P., Tang, C. and Weisenfeld, K., 1987. Self-Organized Criticality: an Explanation of the 1/f Noise, *Phys. Rev. Lett.*, **59**, 381–384.

[10] Bak, P., Tang, C. and Weisenfeld, K., 1988. Self-Organized Criticality, *Phys. Rev. A Gen. Phys.*, **38**, 364.

[11] Bak, P. and Chen, K., 1991. Self-Organized Criticality, Large Interactive Systems Naturally Evolve towards a Critical State ... *Scientfic American*, **264, 1** 26–33.

[12] Bak, P., Chen, K. and Creutz, M., 1989. Self-Organized Criticality in the "Game of Life", *Nature*, **342**, 780–781.

[13] Balakrishnan, A.V., 1976. Applied functional analysis, Springer-Verlag.

[14] Balatoni, J. and Renyi, A., 1956. *Pub. Math. Inst. Hungarian Acad. Sci.*, **1** 9.

[15] Barraclough, D.R. and De Santis, A., 1997, Some Possible Evidence for a Chaotic Geomagnetic Field from Observational Data. *Phys. Earth Plan. Int.*, **99**, 207–220.

[16] Barnsley, M.F., 1988, *Fractals everywhere*, Academic Press Inc.. 394 p.

[17] Beale, R. and Jackson, T., 1994. *Neural Computing: an Introduction*, Bristol, Philadelphia: Inst. of Physics Publ., 240 p.

[18] Beauvais, A., Dubois, J. and Badri, A., 1994. Application d'une analyse fractale à l'étude morphométrique du tracé des cours d'eau : méthode de Richardson, *C. R. Acad. Sci. Paris*, **318, II**, 219–225.

[19] Benoît, E., (Ed.) 1991. *Dynamic Bifurcations*, Lectures Notes in Mathematics, Springer-Verlag, Berlin, **1493**, 219 p.

[20] Bergé, P., Pomeau, Y. et Vidal, C., 1984, *L'ordre dans le Chaos. Vers une Approche Dterministe de la Turbulence.* Hermann, ed. des sciences et des arts, Paris. 353 p.

[21] Blanter, E.M., Schnirman, M.G., Le Mouel, J.L. and Allègre, C.J., 1997, Scaling Laws in Blocks Dynamic Self-Organized Criticality. *Phys. Earth Plan. Int.*, **99**, 295–307.

[22] Bongard, M., Vainzvaig, M., Guberman, Sh., Izvekova, M. and Smirnov, M., 1966. Application of a Self-learning Programme for Identification of the Oil-bearing Layers, *Geology and Geophysics*, **6, II**,

[23] Borizov, B. and Reisner, G., 1974. On the Seismotectonic Catalogue of Caucasus; *Isvestia Rus. Acad. Sci., Phys. Earth*, **N9**,

[24] Borisov, A., Polshakov, M., Poliakov, A., Shukin, Yu., 1975. Usage of Exploration Geophysical Data for Seismic Zoning (Pamir-Tian-Shan region), *Applied Geophysics, Moscow, Nedra*, **N78**,

[25] Bow, S.T., 1992. *Pattern Recognition and Image Pre-Processing*, New York, etc.: Marcel Dekker, 558 p.

[26] Brachman, J., Levesque, H.J. and Retier, R., 1992. *Knowledge Presantation*, Cambridge etc.: A Bradford Book, 408 p.

[27] Briggs, P.L., 1978. Pattern Recognition Applied to Uranium Prospecting, *Ph.D. thesis, M.I.T., Dept. of Earth and Planet. Sciences, USA*,

[28] Bune, V., Toutbovich, I., Borisov, B., Gitis, V., Reisner, G. and Yurkov, E., 1974. Concerning a Method of Identification of Magnitude Dependence of Tectonic Parameters of the Region. *DAN USSR*, **214, N3**

[29] Bunge, M., 1975. Philosophy of Physics, *Moscow, Progress*

[30] Burridge, R. and Knopoff, L., 1967. Model and theoretical seismicity. *Bull. Seism. Soc. Am.*, **57**, 341–371.

[31] Cisternas, A., Gvishiani, A., Godefroy, P., Soloviev, A., Gorshkov, A., Kossobokov, V., Lambert, M., Rantsman, E., Sallantin, J., Soldano, A. and Weber, C., 1985. A Dual Approach to Recognition of Earthquakes Prone-areas in the Western Alps, *Annales Geophysicae*, **3 N2**, 249–270.

[32] Cisternas, A., Gvishiani, A., et al., 1989. The Spitak (Armenia) Earthquake of 7 December 1988: Field Observations, Seismology and Tectonics, *Nature*, **339**, 645–679.

[33] Cortini, M. and Barton, C.C., 1994. Chaos in Geomagnetic Reversal Records: A Comparison between Earth's Magnetic Field Data and Model Disk Dynamo Data, *J. Geophys. Res.*, **99**, B9, 18,021–18033.

[34] Courtillot, V. and Le Mouel, J.L., 1988. Time Variations of the Earth's Magnetic Field: from Daily to Secular, *Ann. Rev. Earth Planet. Sci.*, **16**, 389–476.

[35] Cox, A., 1968. Lengths of Geomagnetic Polarity Intervals, *J. Geophys. Res.*, **73**, 3247–3260.

[36] Cox, E., 1994. *The Fuzzy Systems Handbook: A Practitioner's Guid to Building, Using and Maintening Fuzzy Systems*, Boston, A.P. Professional, 624 p.

[37] Croquette, V., 1987. Deux exemples mécaniques ayant des comportements chaotiques, *Traitement numérique des attracteurs étranges, Éd. CNRS, Paris*, 107–127.

[38] Cundall, P.A., 1971. A Computer Model for Simulating Progressive, Large Scale Movements in Blocky Rock System, *Symp. Int. Soc. Rock Mech., Nancy, 1971.*

[39] Curry, J. and Yorke, J.A., 1977. A Transition from Hopf Bifurcation to Chaos: Computer Experiments with Maps in R2, in *The Structure of Attractors: Dynamical Systems, Lecture Notes in Math.*, **668**, *48, Springer-Verlag.*

[40] Dorr, B.J., 1993. *Machine Translation: a View from the Lexicon*, Cambridge etc.: The MIT Press, 432 p.

[41] Dougherty, E.R. and Laplante, P.A., 1995. *Introduction to Real-time Imaging*, New York: SPIE Optical Engineering Press, 198 p.

[42] Dubois, D. and Prade, H., 1986. *Possibility Theory: an Approach to Computized Processing of Uncertanity* New York, London: Plenum Press, 263 p.

[43] Dubois, J.E. and Gershon, N., 1996. *Modeling Complex Data for Creating Information*, Data and Knowledge in a Changing World, Springer, Codata, 277 p.

[44] Dubois, J.O., 1995. *La Dynamique Non-Linéaire en Physique du Globe*, Masson, Paris, 292 p.

[45] Dubois, J.O. et Cheminée, J.L., 1988. Application d'une Analyse Fractale à l'Étude des Cycles Éruptifs du Piton de la Fournaise (La Réunion) : Modèle d'une Poussière de Cantor, *C. R. Acad. Sci. Paris*, **307, II**, 1723–1729.

[46] Dubois, J.O. and Cheminée, J.L., 1991. Fractal Analysis of Eruptive Activity of Some Basaltic Volcanoes, *J. Volcan. Geotherm. Res.*, **45**, 197–208.

[47] Dubois, J.O. and Nouaili, L., 1989. Quantification of the Fracturing of the Slab using a Fractal Approach, *Earth Plan. Sci. Lett.*, **94**, 97–108.

[48] Dubois, J.O., Chaline, J. et Brunet-Lecomte, P., 1992. Spéciation, Extinction et Attracteurs Étranges, *C. R. Acad. Sci. Paris*, **315**, 1827–1833.

[49] Dubois, J.O. et Cheminée, J.L., 1993. Les Cycles Éruptifs du Piton de la Fournaise: Analyse Fractale, Attracteurs, Aspects Déterministes, *Bull. Soc. Géol. France*, **164, 1**, 3–16.

[50] Dubois, J.O. et Pambrun, C., 1990. Étude de la Distribution des Inversions du Champ Magntique Terrestre entre -165 Ma et l'Actuel (Échelle de Cox). Recherche d'un Attracteur dans le Systme Dynamique qui les Génère. *C. R. Acad. Sci. Paris*, **311, II**, 643–650.

[51] Eliutin, A. and Beriozko, A., 1996. Interactive Earthquakes Catalogs Processing in the Mult-Functional Geo-Information Environment "Attila": How much a Catalog is Needed for Seimic Zonning? *Cahiers du Cent. Europ. Géodyn. et Séismol.*, **12**, 73–82.

[52] Eliutin, A., Morat, P., Pride, S., Le Mouel, J.L. and Gvishiani, A., 1996. Geo-Information Environment for Non-Geographical Data: Processing Spatially Distributed Electrical Signals on Strssed and Deteriorated Rocks Samples, *Cahiers du Cent. Europ. Géodyn. et Séismol.*, **12**,

[53] Ershov, S.V., Malinetskii, G.G. and Ruzmaikin, A.A., 1989. A Generalized Two-Disk Dynamo Model. *Geophys. Astrophys. Fluid Dynamics*, **47**, 251–277.

[54] Falconer, K., 1990. Fractal Geometry: Mathematical Foundations and Applications, *Ed Jhon Wiley and Sons, Chichester*, 288pp.

[55] Farmer, J.D., Ott, E. and Yorke, J.A., 1983. The Dimension of Chaotic Attractors, *Phisica D*, **7**, 153–180.

[56] Fedotov, S., 1965. Distribution of Strong Earthquakes of Kamchatka, Kuril Islands and North-eastern Japan. Questions of Engineering Seismology, *Moscow, Nauka* **10**,

[57] Feigenbaum, M.J., 1978. Quantitative Universality for a Class of Nonlinear Transformations, *J. Statist. Phys.* **19, 1**, 25-52.

[58] Feigenbaum, M.J., 1979. The Universal Metric Properties of Non-linear Transformations, *J. Statist. Phys.* **21, 6**, 669-706.

[59] Firebaugh, M.W., 1989. *Artificial Intelligence. A Knowledge-based Approach* Boston: PWS Kent, 740 p.

[60] Fraser and Swinney, 1986. voir Hongre et al.

[61] Freeman, J.A. and Skapura, D.M., 1992. *Neural Networks: Algorithms, Applications and Programming Techniques*, New York, etc.: Addison Wesley, 412.

[62] Frisch, U. and Parisi, G., 1985. Turbulence and predictability of geophysicalfluids dynamics and climate dynamics, *Proc. Int. School of Phys. "Enrico Fermi", Ed. M. Ghil, North Holland, Amsterdam*, 84.

[63] Froehling et al., 1981.

[64] Gabrielov, A., Keilis-Borok, V., Levshina, T. and Shaposhnikov, V., 1986. Block Model of the Lithosphere Dynamics, *Vichistelnoiya Seismologia, M., Nauka*, **19**, 168–178.

[65] Gelfand, I., 1971. *Linear Algebra* Moscow, Nauka, 212 p.

[66] Gelfand, I., Guberman, Sh., Zhidkov, M., Kalezkaya, M., Keilis-Borok, V. and Ranzman, E., 1973. Transfer of High Seismicity Criteria from Central Asia to Anatolya and Adjacent Regions, *DAN USSR*, **210, N2**,

[67] Gelfand, I., Guberman, Sh., Zhidkov, M., Kalezkaya, M., Keilis-Borok, V., Ranzman, E. and Rotwain, I., 1974. Recognition of Places Where Strong Earthquake May Occur II. Four Regions of Minor Asia and South-Eastern Europe, *Computational Seismology, Moscow, Nauka*, **7**,

[68] Gelfand, I., Guberman, Sh., Zhidkov, M., Keilis-Borok, V. and Ranzman, E. and Rotwain, I., 1974. Recognition of Places where Strong Earthquake may occur III. Case, where the Boundaries of Disjunctive Knots are Unknown, *Computational Seismology, Moscow, Nauka*, **7**, 41–65

[69] Gelfand, I.M., Guberman, Sh.A., Keilis-Borok, V.I., Knopoff, L., Press, F., Rantsman, E.Ya., Rotwain, I.M. and Sadovsky, A.M., 1976. Pattern Recognition Applied to Earthquake Epicenters in California, *J. Phys. Earth and Planet. Int.*, **11**,

[70] Gelfand *et al.* Applications of Pattern Recognition Algorithm "Cora" to Medical Diagnostics

[71] Giarratano, J. and Riley, G., 1994. *Expert Systems: Principles and Programming, 2nd ed.*, Boston: PWS Publ. 644 p.

[72] Goldberg, D.E., 1989. *Genetic Algorithm in Search, Optimization and Machine Learning*, Readin Mass. etc.: Addison Wesley, 412 p.

[73] Gorskhov, A., Caputo, M., Keilis-Borok, V., Ofizerova, E., Ranzman, E. and Rotwain, I., 1979. Recognition of Places where Strong Earthquakes May Occur. IX. Italy, M ¿ 6,0, *Computational Seismology, Moscow, Nauka*, **12**, 3–18.

[74] Gourvitch, V., 1973. To the Theory of Multi-steps Games, *Computational Mathematics and Mathematical Physics*, **13, N6**,

[75] Grassberger, P. and Procaccia, I., 1983. Characterization of Strange Attractors, *Phys. Rev. Lett.*, **50, 5**, 346–349.

[76] Guckenheimer, J., 1986. *Ann. Rev. Fluid Mech.*, **18**, 15–31.

[77] Guckenheimer, J. and Holmes, P., 1983. *Nonlinear Oscillations, Dynamical Systems, and Bifurcations of Vector Fields* Springer ed., New York Berlin Heidelberg and Tokyo.

[78] Gutenberg, B. and Richter, C.F., 1954. *Seismicity of the Earth and Associated Phenomena*, 2nd ed, Princeton University Press.

[79] Gvishiani, A.D., 1982. Time-stability of Strong Earthquake Prone Areas. I. South-Eastern Europe and Minor Asia, *Izvzstia Akademii Nauka USSR; Physics of the Earth*, **8**

[80] Gvishiani, A.D., Gorshkov, A. and Kossobokov, V., 1987. Recognition of Seismicaly Active Zones in Pyrénés, *DAN USSR*, **292, N1**, 56–59.

[81] Gvishiani, A.D., Gorshkov, A., Zhidkov, V. Ranzman, E. and Troussov, A., 1987. Recognition of Places where Strong Earthquakes May Occur, XV, Morphostructural Knots of the Great Caucasus $M \geq 5.5$, *Computotianal Seismology Moscow, Nauka*, 136–148.

[82] Gvishiani, A.D., Gorshkov, A., Ranzman, E., Cisternas, A. and Soloviev, A., 1988. *Recognition of Earthquake Prone-areas in Regions of Moderate Seismicity*, Moscow, Nauka, 176 p.

[83] Gvishiani, A.D. and Gurvitch, V.A., 1987. Calculation of Lithosphere Blocks Equilibrium by Convex Programming Methods, *XIX General Assembly IUGG, Vancouver, Canada, Abstracts*, **V1**, 39.

[84] Gvishiani, A.D. and Gurvitch, V.A., 1992. *Dynamical Classification Problems and Convex Programming in Applications*, Moscow, Nauka, 355 p.

[85] Gvishiani, A.D. and Gourvitch, V., 1982. Time-stability of Earthquake Prone Areas Recognition II. Eastern Part of Central Asia, *Izvestia AN USSR, Physics of the Earth*, **N9**, 30–38.

[86] Gvishiani, A.D. and Gourvitch, V., 1983. Dual Systems and Sets and its Applications, *Izvestia AN USSR, Technical Cybernetics*, **N4**, 31–39.

[87] Gvishiani, A.D. and Gourvitch, V.A. and Raszvetaev, A., 1985. Dynamical Problems of Pattern Recognition III. Study of Stability of the Recognition of Places of Strongest Earthquakes of Pacific Belt, *Computational Seismology, Moscow, Nauka*, **18**,

[88] Gvishiani, A.D. and Gourvitch, V. and Raszvetaev, A., 1986. Estimation of Seismic Fracturing by Pattern Recognition Technique, *Computational Seismology, Moscow, Nauka*, **19**, .

[89] Gvishiani, A.D., Zelevinsky, A., Keilis-Borok, V. and Kossobokov, V., 1978. Study of the Strongest Earthquake Prone-areas in the Pacific Belt, *Izvestia AN USSR, Physics of the Earth*, **N9**, 31–42.

[90] Gvishiani, A.D. and Kossobokov, V., 1981. On Evaluation of Reliability of Results of Pattern Recognition of Strong Earthquake-Prone Areas, *Izvestia AN USSR, Physics of the Earth*, **N2**, 21–36

[91] Gvishiani, A.D. and Soloviev, A., 1982. Concerning the Solution of the Problem of Recognition of Strong Earthquake Prone-areas on the Pacific Coast of South America, *Izvestia AN USSR, Physics of the Earth*, **N1**, 86–87.

[92] Gvishiani, A., Gorshkov, A., Kossobokov, V., Cisternas, A., Philip, H. and Weber, C., 1987. Identification of Seismicaly Dangerous Zones in the Pyrenees, *Annales Geophysicae*, **5 (6)**, 681–690.

[93] Gvishiani, A.D., Zhizhin, M.N., Mikoyan, A.N., Bonnin, J. and Mohammadioun, B., 1995. Syntactic Analysis of Waveforms from the World-wide Strong Motion Database, *European Seismic Design Practice, Rotterdam, Balkema, Brookfied*, 557–564.

[94] Hall, 1970. Combinatorics, *Moscow, Mir*,

[95] Halsey, T.C. and Jensen, M.H., 1986. Spectra of Scaling Indices for Fractal Measures/ Theory and Experiment, *Physica D*, **23**, 112-117.

[96] Halsey, T.C., Jensens, M.H., Kadanoff, L.P., Procaccia, I. and Shraiman, B.I., 1986. Fractal measures and their Singularties: The Characterization of Strange Sets, *Physical Review A*, **33, 2**, 1141–1151.

[97] Haykin, S. and Griffin, J., 1994. *Neural networks. A comprehensive foundation*, New York, etc.: Macmillan Publishing Company, 696 p.

[98] Held, G.A., Solina, D.H., Keane, D.T., Haag, W.J., Horn, P.M. and Grinstein, G., 1990. Experimental Study of Critical-Mass Fluctuations in an Evolving Sandpile, *Phys. Rev. Lett.* **65, 9**, 1120–1123.

[99] Hentschel, H.G.E. and Procaccia, I., 1983. The Infinite Number of Generalized Dimensions of Fractals ansd Strange Attractors, *Physica 8D*, 435–444.

[100] Hertz, J., Krogh, A. and Palmer, R.G., 1994. *Introduction to the Theory of Neural Computation*, Reading Mass. etc.: Addison-Wesley, 327 p.

[101] Hénon, M., 1976. A Two-dimensional Mapping with a Strange Attractor, *Communications in Math. Phys.*, **565**, 29.

[102] Hide, R., 1995. Structural Instability of the Rikitake Disk Dynamo, *Geophys. Res. Lett.*, **22**, 1057–1059.

[103] Hide, R., Skeldon, A.C. and Acheson, D.J., 1996. A study of two novel self-exciting single-disk homopolar dynamos: theory, *Proc. R. Soc. Lond. A.*, **452**, 1369–1395.

[104] Holland, J.H., 1992. *Adaptation in Natural and Artificial Systems* Cambridge: The MIT Press, 211.

[105] Homeinuk, Yu., Shukin, Yu., Firsova, D. and Filippova, G., 1978. Method of Evaluation of Probability of the Strongest Earthquakes by Geological Geophysical Data Analysis, *Moscow, Nauka*,

[106] Hongre, L., Zhizhin, M. and Dubois, J., 1995. Chaotic Characteristics of the Magnetic Field : Fractal Dimensions and Lyapunov Exponents, *Cahiers du Centre Européen de Géodynamique et de Sismologie*, **9**, 79-94.

[107] Hutchinson, A., 1995. *Algorithm Learning* Oxford: Clarendon Press, 436 p.

[108] Hwa, T. and Kardar, M., 1989. Dissipative Transport in an Open System: anInvestigation of Self-0rganized Criticality, *Phys. Rev. Lett.*, **62, 16**, 1813–1816.

[109] Ito, K., 1980. Chaos in the Rikitake Two-disc Dynamo System, *Earth Plan. Sc. Lett.*, **51**, 451–456.

[110] Jones-Cecil, M., Wheeler, R.L. and Dewey, J.W., 1981. Pattern Recognition Program Modified and Applied to Southerneastern Ynited States Seismicity, *U.S. Geol. Survey Open-file Report*, 81–195.

[111] Kadanoff, L.P., Nagel, S.R., Wu, L. and Zhou, S.M., 1989. Scaling and Universality in Avalanches, *Phys. Rev. A*, **39**, 6524–6532.

[112] Kaplan, J. and Yorke, J., 1978. Functional differential equations and the approximation of fixed points, *Proceedings, Bonn, July 1978, Lectures notes in Math.*, **730**, *H.O. Peitgen and H.O. Walther, eds., Springer, Berlin, 1978*, 228.

[113] Kapral, R. and Mandel, P., 1985. Bifurcation structure of nonautonomous quadratic map, *Phys. Rev. A*, **32**, 1076–1080.

[114] Kartalopoulos, S.V., 1996. *Understanding Neural Networks and Fuzzy Logic: Basic Concepts and Applications* New York etc.: IEEE Press, 205 p.

[115] Keilis-Borok, V.J., 1990. The Lithosphere of the Earth as a Non-linear System with Implications for Earthquake Prediction, *Rev. Geophys.*, **28**, 19-34.

[116] Kirillov, A.A. and Gvishiani, A.D, 1982. *Theorems and Problems in Functional Analysis*, Springer-Verlag, New York Inc., 347 p.

[117] Kolmogorov, A. and Fomin, C., 1972. *Elements of Theory of Functions "and Functional Analysis"* Nauka, Moscow, 496 p.

[118] Kosko, B., 1992. *Neural networks and Fuzzy Systems: A Dynamical System Approach to Machine Intelligence*, Englewood Cliffs: Prentice Hall, 452 p.

[119] Kosko, B., 1994. *Fuzzy Thinking. The New Science of Fuzzy Logic*, Glasgow: Harper Collins, 318 p.

[120] Landau, L. et Lifchitz, E., 1967. *Théorie de l'Élasticité.* Éditions MIR, Moscou.

[121] Landau, L. et Lifchitz, E., 1971. *Mécanique des fluides.* Éditions MIR, Moscou. 669 p.

[122] Ledésert, B., Dubois, J., Velde, B., Meunier, A., Genter, A. and Badri, A., 1993. Geometrical and Fractal Analysis of Three-dimensional Hydrothermal Vein Network in a Fractured Granite, *J. Volcanol.Geotherm. Res.*, **56**, 267–280.

[123] Ledésert, B., Dubois, J., Genter, A. and Meunier, A., 1993. Fractal Analysis of Fractures applied to Soultz-sous-Forets Hot Dry Rock Geothermal Program, *J. Volcanol.Geotherm. Res.*, **57**, 1–17.

[124] Levy, S., 1993. *Artificial Life. The Quest of a New Creation* London, etc.: Penguin books, 390 p.

[125] Lorenz, E.N., 1963. Deterministic Non-periodic Flow, *J. Atmosph. Sci.*, **20**, 130.

[126] Mackey, M.C., and Glass, L., 1977. Oscillation and Chaos in Physiological Control Systems, *Science*, **197**, 287–289.

[127] Malraison, B., Atten, P., Bergé, P. and Dubois, M., 1983. Dimension d'attracteurs étranges : une détermination expérimentale en régime chaotique de deux systèmes convectifs, *C. R. Acad. Sci. Paris*, **297**, 209–214.

[128] Mandelbrot, B.B., 1977. *Fractal-Form, Chance and Dimension*, Freeman, San Francisco.

[129] Mané, R., 1981. Dynamical Systems and Turbulence, Warwik 1980, *Lecture Notes in Mathematics, Springer, Berlin*, **898**, 230.

[130] Marks II, R.J., 1994. *Fuzzy Logic Technology and Applications*, New York (IEEE Technology Update Series), 575 p.

[131] Marzocchi, W., 1997. Missing Reversals in the geomagnetic Polarity Timescale: Their Influence on the Analysis and in Constraining the Process the Generates Geomagnetic Reversals, *J. Geophys. Res.*, **102**, B3, 5157-5171.

[132] Marzocchi, W. and Mulargia, F., 1992. The periodicity of geomagnetic reversals, *Phys. Earth Planet. Int.*, **73**, 222–228.

[133] Marzocchi, W., Gonzato, G. and Mulargia, F., 1995. Rikitake's Geodynamo Model Analysed in terms of Classical Time Series Statistics, *Phys. Earth Planet. Int.*, **88**, 83–88.

[134] McNeil, D. and Freiberger, P., 1994. *Fuzzy Logic. The revolutionary Computer Technology that is Changing our World*, New York, etc.,: Simon and Schuster, 319 p.

[135] Mikoyan, A, Burstev, A., Gvishiani, A. and Zhizhin, M., 1997. EMSC Strong Motion Database: WWW Interface, *EMSC Newsletter*,

[136] Mora, P., 1989a, *Elastic Wavefield Inversion of Reflection and Transmission Data*, **99**, 1211-1233.

[137] Mora, P., 1989b, *Nonlinear Elastic Wavefield Inversiof Multi-offset Seismic Data*, **99**, 1222-2244.

[138] Orchard, G.A. and Phillips, W.A., 1991. *Neural Computation: A Beginner's Guide*, Hove etc.: Lawrence Erlbaum, 141 p.

[139] Osedelets, V., 1967. Multiplicative ergotic theorem. Lyapunov characteristics numbers for dynamic systems, *Trans. Moscow Mathem. Soc.*, **19**, 197–231.

[140] Packard et al., 1980.

[141] Panteleyev, A.N., Dubois, J. and Diament, M., 1995. Roughness of the Global Altimetric Geoid, *Cahiers du Centre Erňopéen de Géodynamique et de Séismologie*, **9**, 43–77.

[142] Philip, H., Cisternas, A., Gvishiani, A. and Gorshkov, A., 1989. The Caucasus: an Actual Example of the Initial Stages of Continental Collision, *Tectonophysics*, **161**, 1–21.

[143] Mira, C., 1987. Généralités surl'outil de Récurrence. Ed. CNRS, Paris.

[144] Philip, H., Roghozin, E., Cisternas, A., Bousquet, J.C., Borisov, B. and Karakhanian, A., 1992. The Armenian Earthquake of 1988 December 7: Faulting and Folding, Neotectonics and Paleosismicity, *Geophys. J. Int.*, **110** 141–158.

[145] Pinsker, I., 1973. Evaluation of the Method of Learning and Learning Set. Modeling and Automated Analysis of Electrocardiograms, *Moscow, Nauka*,

[146] Rabiner, L. and Juang, B.H., 1993. *Fundamentals of Speech Tecognition*, Englewood Cliffs: Prentice Hall, 505 p.

[147] Raszvetaev, A., 1984. Local Stability of Seismic Prediction and Cluster Analysis, *Computational Seismology, Moscow, Nauka*, **17**, 67–69.

[148] Regional Catalogue of Earthquakes. International Seismological Center, *Newberry, Edinbourg, 1964-1982*, **V. 1-18**,

[149] Rice, J.R., 1979. Theory of precursory processes in the inception of earthquake rupture, *Gerlands Butrage zur Geophysik*, **88, N2**,

[150] Rikitake, T., 1958. Oscillations of a system of disk dynamos, *Proc. Camb. Phil. Soc.*, **54**, 89–105.

[151] Rössler, O.E., An Equation for Continuous Chaos, 1976. *Phys. Lett. A* **57**, 397–398.

[152] Rossler, O.E., 1978. Continuous chaos, in Sygernetics - A Workshop, H. Haken, ed., Springer, New York, 184 p.

[153] Ruelle, D. and Takens, F., 1971. On the nature of the turbulence, *Com. Mathem. Phys.*, **20**, 167–192, **23**, 343–344.

[154] Ruelle, D., 1987. *Chaotic Evolution of Strange Attractor*, Cambridge Universty Press.

[155] Ruelle, D., 1990. voir Hongre et al

[156] Ruelle, D., 1994. voir bifurcations

[157] Russ, J.C., 1995. *The Image Processing Handbook, 2nd ed.*, Boca Raton etc.: CRC Press, 674 p.

[158] Schalkoff, R.J., 1992. *Pattern Recognition: Statistical, Structural and Neural Approches*, New York: John Wiley, 364 p.

[159] Shapiro, S.,C., 1990. *Encyclopedia of Artificial Intelligence, in 2 vol.*, New York etc.: Wiley, vol.1, 679 p.

[160] Shapiro, S.,C., 1990. *Encyclopedia of Artificial Intelligence, in 2 vol.*, New York etc.: Wiley, vol.2, 680–1219.

[161] Shebalin, P., Girardin, N., Rotwain, I., Keilis-Borok, V. and Dubois, J., 1996. Local Overturn of Active and Non-active Seismic Zones as a Precursor of Large earthquakes in the Lesser Antillean Arc, *Physics of the Earth and Planetary Interiors*, **97**, 163–175.

[162] Shilov, G., 1952. *Introduction to linear spaces theory* GITTL, Moscow, 312 p.

[163] Shinbrot, T., Grebogi, C., Ott, E. and Yorke, J.A., 1993. Using Small Perturbations to Control Chaos, *Nature*, **363**, 411–417.

[164] Simpson, P.K., 1996. *Neural Networks Theory, Technology and Applications*, New York: (IEEE Technology Update Series) 943 p.

[165] Smalley, R.F., Chatelain, J.L., Turcotte, D.L. and Prévot, L., 1987. A fractal approach to the clustering of earthquakes: application to the seismicity of the New Hebrides, *Bull. Seismol. Soc. Am.*, **77**, 4, 1368–1381.

[166] Sobolev, G., 1981. Modeling of the Process of Preparation and Precursors of Earthquakes. Physical Properties of the Mountains Rocks, *Tashkent*,

[167] Sornette, D., 1993. Physique Statistique, les Phénomènes Critiques Auto-Organisés, in: *Images de la Physique 1993*, 9–17.

[168] Sornette, A. and Sornette, D., 1989. Self-Organised Criticality and Earthquakes, *Europhys. Lett.*, **9**, 197–202.

[169] Sornette, D., Davy, P. and Sornette, A., 1990. Structuration of the Lithosphere in Plate Tectonics as a Self-Organized Criticality Phenomenon, *J. Geophys. Res.*, **95**, 17353–17361.

[170] Sornette, A., Dubois, J., Cheminée, J.L. and Sornette, D., 1991. Are Sequences of Volcanic Eruptions Deterministically Chaotic ? *J. Geophys. Res.*, **96**, 11,931-11,945.

[171] Sugihara, G. and May, R.M., 1990. Non-linear Forecasting as a Way of Distinguishing Chaos from Measurement Error in Times Series, *Nature*, **344**, 734-741.

[172] Sukmono, S., Zen, M.T., Kadir, W.G.A., Hendrajaya, L., Santoso, D. and Dubois, J., 1996. Fractal Geometry of the Sumatra Active Fault System and its Geodynamical Implications, *J. Geodynamics*, **22 1/2**, 1–9.

[173] Takens, F., 1981. Detecting strange attractors in turbulence, *Lecture Notes in Mathematics, Springer, Berlin*, **898**, 366.

[174] Terano, T., Asai, K. and Sugeno, M., 1987. *Fuzzy Systems Theory and its Applications*, Boston, etc.: Academic Press, 268.

[175] Tu, D. and Gonzales, P., 1978. Principals of Pattern Recognition, *Moscow, Mir*

[176] Turcotte, D.L., 1992. *Fractals and Chaos in Geology and Geophysics*, Cambridge University Press, 221 p.

[177] Vainzvaig, M., 1973. Algorithms of Pattern Recognition with Learning "Cora" Algorithms of a Pattern Recognition Learning, *Moscow, Sov. Radio*,

[178] Weber, C., Gvishiani, A., Godefroy, P., Gorshkov, A., Kossobokov, V., Lambert, J., Ranzman, E., Sallantin, J., Soldano, A., Cisternas, A. and Soloviev, A., 1986. Recognition of Places where Strong Earthquakes May Occur XII. Two Approaches to Recognition of Strong Earthquakes in Western Alps, *Computational Seismology, Moscow, Nauka*, **19**, 132–154

[179] Weber, C., Gvishiani, A., Godefroy, P., Lambert, J., Ranzman, Soloviev, A. and Trusov, A., 1986. Recognition of Places where Strong Earthquakes May Occur XIII. Neotectonic Scheme of Western Alps; $M \geq 5,0$, *Computational Seismology, Moscow, Nauka*, **19**, 82–94.

[180] Wolf, A., Swift, J.B., Swinney, H.L. and Vastano, J.A., 1985. Determining Lyapunov exponents from a time series, *Physica D*, **16**, 285–317.

[181] Yaglom, I., 1980. Boolean Structure and its Models, *Moscow, Sov. Radio*,

[182] Zax, Sh., 1975. Theory of Statistical Conclusions, *Moscow Mir*,

[183] Zhizhin, M., Gvishiani, A., Bottard, S., Mohammadioun, B. and Bonnin, J., 1992. Clasification of Strong Motion Waveforms from Different Geological Regions using Syntactic Pattern Recognition Scheme, *Cahiers Centre Europ. Géodyn. Séism.*, **6**, 33–42.

[184] Zhizhin, M., Gvishiani, A., Bonnin, J., Madariaga, R., Mohammadioun, B. and Rouland, D., 1995. Syntactic Pattern Recognition Scheme (SPARS) Applied to Seismological Waveform Analysis, *Cahiers Centre Europ. Géodyn. Séism.*, **9**, 17–26.

[185] Zhizhin, M., Gvishiani, A., Mikoyan, A.N., Bonnin, J. and Mohammadioun, B., 1996. Syntactic Pattern Recognition in Strong Motion Data Bank, *Cahiers Centre Europ. Géodyn. Séism.*, **12**, 271–281.

[186] Zhizhin, M., Bataglia, J., Dubois, J. and Gvishiani, A., 1996. Application of Dynamic Programming for the Reconstruction of Isochrons along the Mid-Atlantic Ridge, *Cahiers du Centre Europ. de Géodyn. et Séismol.*, **12**, 151–159.

# Part VI

# Index

# Index

# G



# H



# I



# J



# K



# L



# M

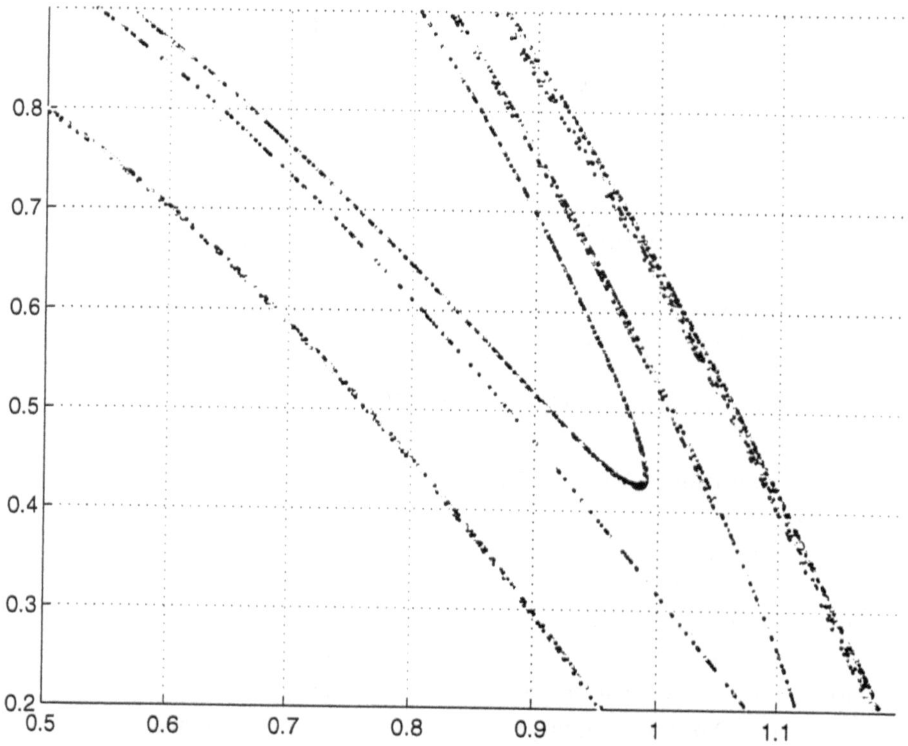


Structure of Hénon attractor - On this enlargement the self similarity in the structure is visible (20 000 points, after David Aubert, 1997).

# Colour Plates

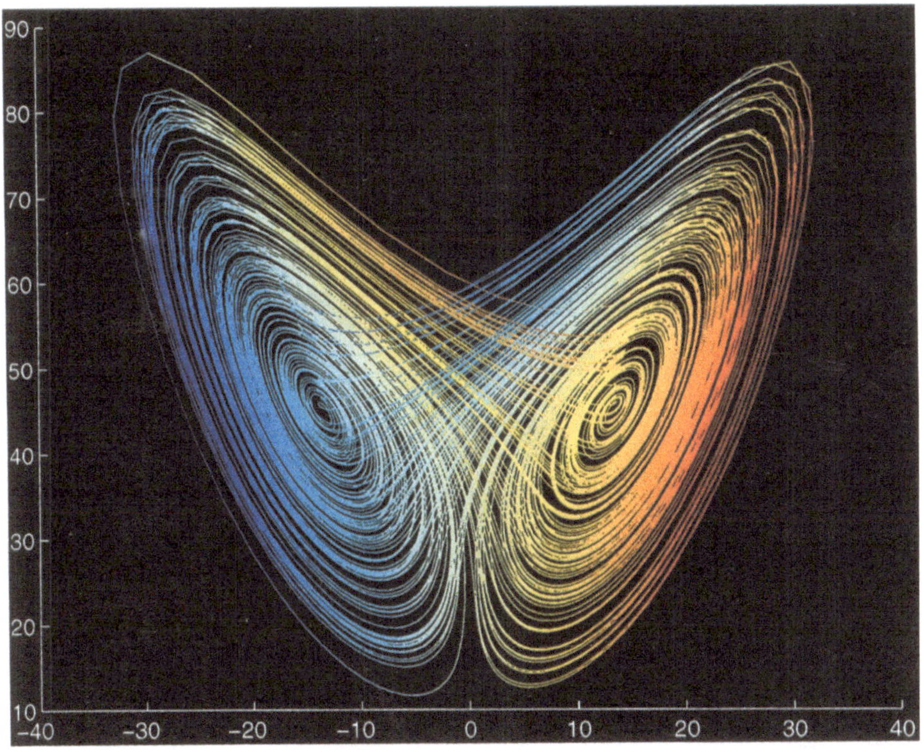


The Lorenz Attractor – The parameters (see the text) are: $Pr = 10$ ; $b = 8/3$; $r = 45.92$. The perspective representation in the phase space shows 10,000 points discrete phase trajectories. The colours are time dependent, from blue at the beginning to red at the end of the series.
(after David AUBERT, 1997).

**Earthquake-prone areas in the Great Caucasus (M ⩾ 6.5).**

**Monitoring of epicenters after publication of the results in 1986.**

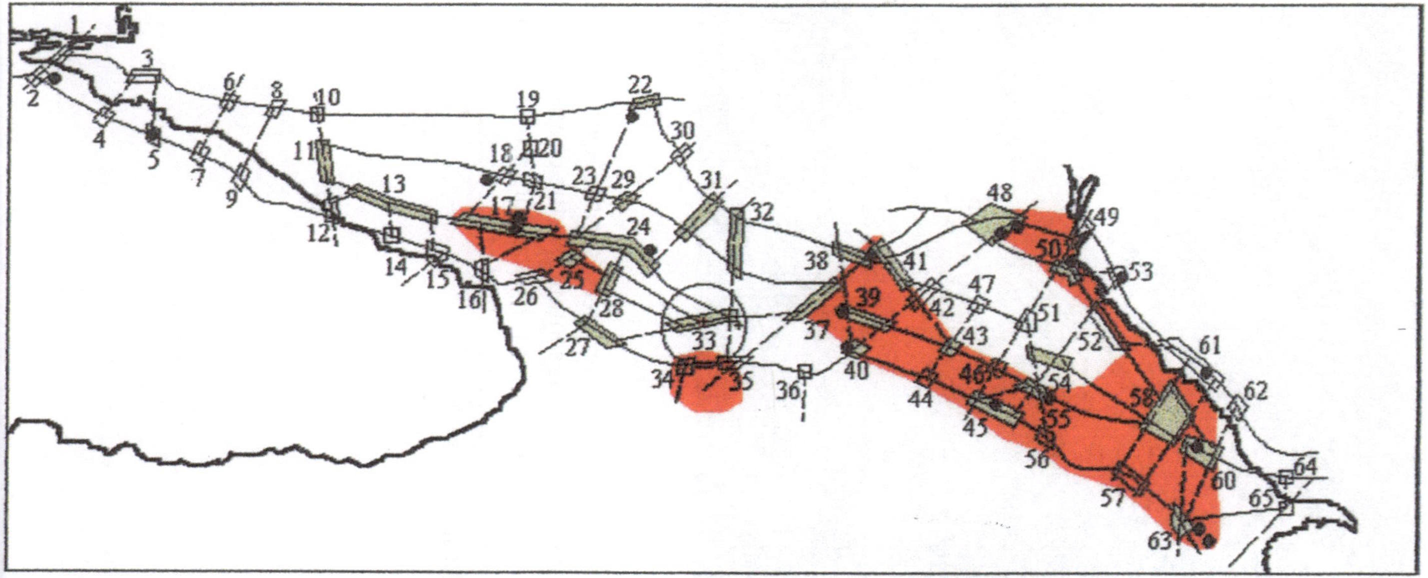

Earthquake-prone areas in the Great Caucasus. Red colour represents zones where intensity 8 is possible in a seismic zoning mapping. Green are the Knots recognized as prone for earthquakes with $M \geq 6.5$. The Racha-Java earthquake (Georgia) in segment 33, $M = 6.3$ occured in 1991 five years after this publication (see page 6).

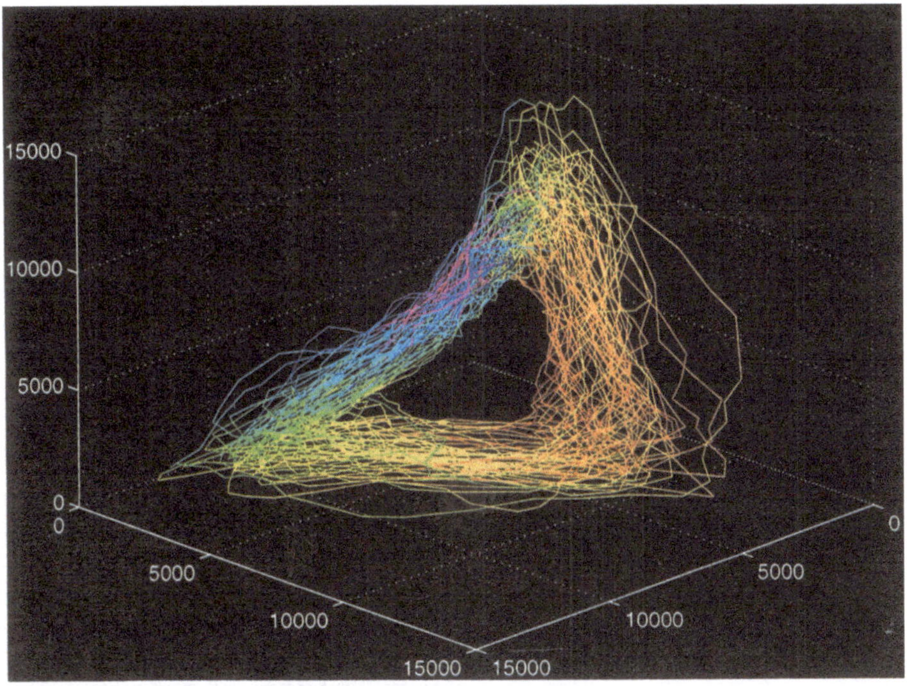


The Hénon Attractor – The parameters (see the text) are: $\alpha = 1.4$ ; $\beta = 0.3$. The representation in the phase space (the plane X, Y of the mapping) shows 20,000 points the colour of which depending on time, blue to red (after David AUBERT, 1997).

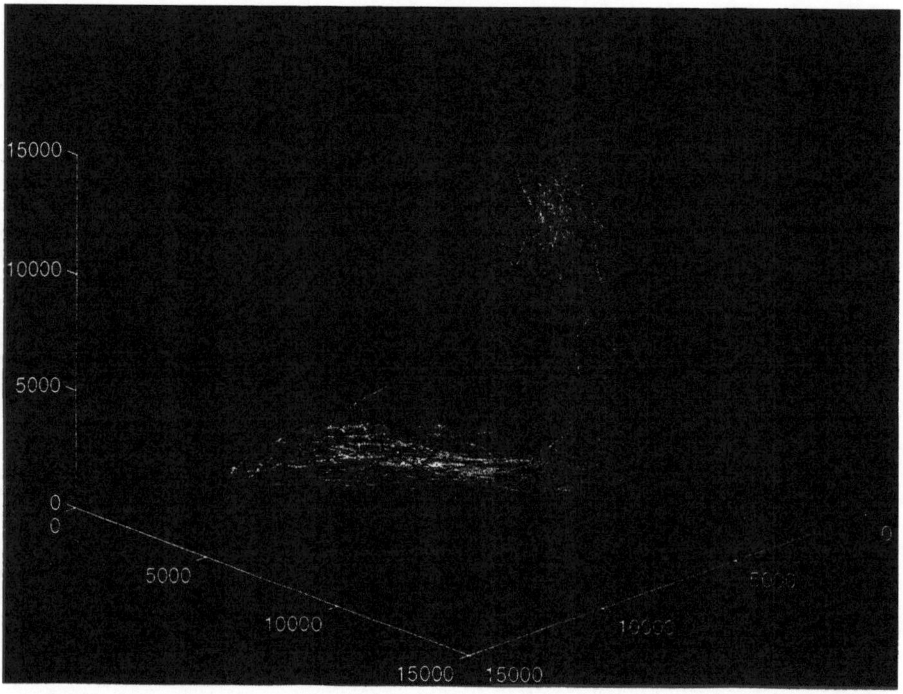


A natural phenomenon attractor – Here is represented in its phase space the Oubangui river discharge during a 61 years period. The 22,000 points of the phase trajectories give a geometry very similar to the Rössler attractor geometry (after David AUBERT, 1997).

# CODATA

CODATA, Committee on Data for Science and Technology, an Association comprised of national committees and scientific unions, is an Interdisciplinary Scientific Committee of the International Council of Scientific Unions (ICSU). It was set up in January 1966 and established in Paris, France and is concerned with the organization, management, quality control, reliability, and dissemination of data and data bases from all scientific and technical disciplines.

Its subject matter scope includes data from broadly varying horizons, ranging from the physical, biological, geological and astronomical sciences to engineering and social science data in complex integrations, e.g. the environment, sustainable development, quality control in industry and large society information systems. For data communication and information transfer in the light of the present Information Revolution, CODATA is centering its efforts on network methodologies and knowledge production through knowledge sharing with the scientific and technical communities.

## Volumes in the Series
## Data and Knowledge in a Changing World

**The Information Revolution:**
**Impact on Science and Technology**
1996, 50 Figures, 17 Tables, 274 p.
Edited by J.-E. Dubois and N. Gershon
ISBN 3-540-60855-9

**Modeling Complex Data for Creating Information**
1996, 84 Figures, 12 Tables, 278 p.
Edited by J.-E. Dubois and N. Gershon
ISBN 3-540-61069-3

**Industrial Information and Design Issues**
1996, 102 Figures, 8 Tables, 294 p.
Edited by J.-E. Dubois and N. Gershon
ISBN 3-540-61457-5

**Geosciences and Water Resources:**
**Environmental Data Modeling**
1997, 95 Figures, 35 Tables, 340 p.
Edited by C. Bardinet and J.-J. Royer
ISBN 3-540-61947-X

**Dynamic Systems and Dynamic Classification Problems**
**with Geophysical Applications**
1998, 76 Figures, 256 p.
by J. O. Dubois and A. Gvishiani
ISBN 3-540-64238-2

**Thermodynamic Modeling and Materials Data Engineering**
1997, 38 Figures, 96 Tables, 340 p.
by J.-P. Caliste, A. Truyol and J. Westbrook (Eds.)
ISBN 3-540-64445-8
*(in preparation)*